A Boy's Own

OFFSHORE

Adventure

by

Brian Page

Published in Scotland and the United Kingdom in 2007
by PlashMill Press

First Edition

A CIP catalogue record for this book is available
from the British Library.

ISBN-13: 978-0-9554535-1-9

Printed and bound by Robertson Printers, Forfar

PlashMill Press
The Plash Mill
Friockheim
Angus DD11 4SH

This book is dedicated to my wife Bobbie who, while I was away playing with other intellectual giants, single-handedly managed a full time lectureship and brought up three children to be decent human beings. My love and everlasting thanks.

Brian Page

One of the symptoms of an approaching nervous breakdown is the belief that one's work is terribly important.

Bertrand Russell (1872 – 1970)

Contents

1

Waving—Not Drowning

It was a lovely afternoon, some one hundred miles out in the North Sea on one of our Production Platforms. Bert, Liam and I were leaning over the rail, taking a well-earned break from dismantling things, and our conversation turned to the size of waves. Contrary to popular belief we didn't always talk about the relative merits of Sophia Loren's womanly bits; mostly—yes, but not always, and especially not in daylight.

The main thing here is not to confuse our North Sea waves with those upon which blond gods in rubber suits stand on ironing boards and crash into rocks in Cornwall. Our waves were deeply masculine, about a mile long and, having come unhindered from somewhere in the Arctic Circle, were not about to let a silly little oil platform get in their way.

Anyway, as we watched the waves surge towards us, we tried to guess their height, as an intellectual exercise—consisting mainly of ill-founded pronouncements followed by a chorus of derision, as our scientific estimates of the actual height, based on staring at the water, ranged wildly from thirty to fifty feet.

Liam said, 'Don't forget, we're trying to estimate the total height of the wave, which is the difference in height between the crest and the trough.' Liam was prone to making obvious statements like this; we called them his *Reader's Digest* moments.

Our frustration, in this search for precision, was summed up by Bert who said, 'The problem is, we're looking down on the sea from about eighty feet up and to be accurate we would need to be lower down.' You can see why we were called engineers.

As ever, Liam had the answer. 'I know where we can get under the deck and drop down much closer to sea level.'

Being nominally in charge, I really should have known better, but the idea seemed innocent enough and was certainly better than taking things apart to see how they worked—or not, as the case may be.

So, with Liam in the lead, off we went to the far end of the platform where, on lifting up a cover in the grating, we saw a vertical ladder welded to one of the giant legs.

Down we went until we reached a narrow walkway running north and south between two of the legs. Suddenly the water looked a whole lot nearer. Gosh, what excitement, we could watch the waves approaching and estimating their height became considerably easier.

This was why we had come to work offshore, to be all alone in the wild elements, about sixty feet under the deck in the company of fine comrades. What more could we ask for? Both Bert and I were full of praise for Liam, which turned out to be somewhat premature and was firmly retracted later.

However, here we were, close to the might of nature and loving every minute of it, until Bert spoiled everything by saying, 'Of course you know that every seventh wave is bigger than the others.'

Liam looked at him with something approaching pity and replied, 'Where the hell did you get that rubbish from? And how would you know which wave was the first one to count from anyway?'

Liam really hated it when Bert or I seemed to know more technical things than he did.

Bert looked suitably offended and retorted, 'Well that's what my Granny used to tell us when we went to the beach in Aberdeen.'

'Stop it, children,' I said in one of my rare managerial moments. 'Let's just enjoy the power of the waves, which is what we have climbed down umpteen feet of ladders to see.'

Suitably chastened, we all turned northwards to watch the onrushing waves and quickly reverted to three little boys jumping up and down, laughing at each other, pointing at the waves as they raced towards us and whooshed under the fragile steel walkway—yes, this was a whole lot better than working in a coal mine.

Then, during a lull in the excitement, Bert suddenly piped up, 'Bloody hell; look at the size of that thing coming our way.'

We looked up and sure enough a massive wall of water about the size of a multi-storey car park was coming straight for us. As we stared in mounting horror, it crashed into one of the platform legs, reared up and continued southwards like a mad express train.

At this point, I found myself staring into two pairs of wide-open eyes as my brave soldiers tried to hide inside my overalls. There then followed a brief dance, reminiscent of one of those insect mating rituals you see on David Attenborough programmes, as each of us tried to push the others to breast the onslaught.

However, a moment later and despite our futile and cowardly attempts at self-preservation, a considerable amount of the North Sea landed on squarely top of us.

As we clung onto the handrails, each other and our (own) private parts, a number of hitherto unknown facts emerged—not only were the waves bigger down here, they were freezing cold, very heavy, very wet and filled boots, pockets and every known orifice in a millisecond.

Bert, who had started wearing his chinstrap casually around the rim of his safety helmet, Clint Eastwood style, found the helmet had been swept off his head and was now, presumably, well on its way to France.

There then followed a most embarrassing period when *esprit de corps* gave way to a good deal of pushing and shoving in our pathetic attempts to get back up the ladder before Liam—and certainly before another of nature's wonders hit us.

Bert and I were sustained during the climb by cursing Liam without either drawing breath or repeating a single swear word, in what turned out to be a vast and impressive repertoire.

But the most embarrassing thing of all was the need to sneak back into the accommodation without dripping seawater all over the lino-tiles—preferably without being seen either.

Bert, who was ever the worrier, said, 'What the hell are we going to say if anyone asks why we're soaking wet?'

Liam cheerfully replied, 'We'll casually mention that we've just tested the deluge system in the well-head module and it works fine.'

However I could see a flaw in this. 'But if we were testing a system which was designed to flood the equivalent of an aircraft hanger in about four seconds, wouldn't we be outside it?'

Bert concurred. 'And won't people be suspicious if they go into the well-head and see it's bone-dry?'

To which I had to add, 'Won't someone in the Control Room notice that the fire pump hasn't been switched on? We would need that to test the deluge system.'

Liam, who hated us picking holes in what he considered to be perfect logic, finally snapped. 'You two ungrateful sods can just stop bloody nit-picking,' he growled. 'I'm away to get changed.' And off he squelched.

Liam always, mistakenly, thought he had an ability to maintain dignity under duress—or something equally puerile.

I, however, was left with a distinct feeling that were he ever to mention the fun to be had skydiving, I should ask for a transfer, or better still, throw him overboard.

2

A Disclaimer

I suppose that 'Being in the right place at the right time' is terribly important if you are a trapeze artist. In my case, entering the previously unknown world of oil exploration and production might best be defined as fortuitous.

I joined in the early 1970's, at a time when the UK offshore adventure was just gathering pace and as a result, met up with what the equal opportunities police now call 'a diverse cross-section' of humanity, consisting of designers, constructors, commissioners, operators, maintainers and assorted supporting players.

After a short time, I realised that most of us had one thing in common; although no one really knew what on earth was going on, we all pretended we did.

As I am now older than I was then, I have been encouraged to set down my recollections of the time spent in the development and operation of various offshore production platforms and, as this is a series of personal reminiscences, I have made no attempt to ensure the events follow a strict chronological order.

Sadly, this would be impossible anyway, as I have now reached that enviable time of life where I can readily recall that my first infant school teacher was a dumpy, motherly lady named Mrs Corrigan, but am often unable to remember where I've left my wallet.

I should also emphasise that although names have been changed to protect reputations and avoid libel action, all the recollections are true and in the main unembellished. As they say about many of the things that happen in life, 'you couldn't make it up'.

You may also notice on reading what follows, an unfortunate tendency to refer to films from time to time. I can't help it; I was brought up during the great movie era and was one of those sad people who used to remain in an empty cinema to read the credits at the end of every film.

Finally, I like to think that one day my grandchildren might read this book, as a change from I-Podding, and be amazed to realise that their granddad was once a 'North Sea Tiger.' Sad to relate, he never completely made the grade, having lacked the courage to wear either an earring or medallion in public.

3

Life in the Last Century

In 1975 when my great offshore adventure began, Scotland had its very own pop group called 'The Bay City Rollers.' The Tartan-clad popsters reduced both teenage girls and music lovers to tears—for very different reasons.

The most popular series on TV was 'The Sweeney.' Hard-drinking Flying Squad coppers Regan (John Thaw) and Carter (Dennis Waterman) chased villains around London in a brown Ford Granada, shouting 'Shut it!' and 'Nick him George' in place of a detailed script. The show became notorious because the cops were almost as bent as the robbers, and the bad guys quite often got away with it.

Not much change there then.

A ten-year-old boy was sent home from a London junior school for the heinous crime of turning up for lessons in long trousers. 'Only boys of 12 and over were allowed to wear the regulation grey worsted,' said the headmaster in a note to the youngster's parents.

There were two attempts in the same month to assassinate Gerald Ford, the U.S. President—and we think terrorism has just been invented. Incidentally, this was a man about whom it was said 'He couldn't walk and chew gum at the same time.'

John Lennon won a long battle to be allowed to stay in the U.S.A. In view of what happened later, it was probably not one of his better decisions.

Jim'll Fix It was a very popular reality TV programme; it is reckoned that the host, Sir Jimmy Savile, received some 350,000 requests a year, and not all of them asking for him to be removed from our screens.

The Conservative Party chose the mad woman Thatcher as its new Leader. She was the first woman to head a British political party and rejected suggestions of great celebrations. She said: 'Good heavens no, there's far too much work to be done.' And when she was finished there was nothing left to celebrate.

Colour TV was officially introduced in Australia—now isn't that interesting.

4

Production Platforms

Just in case anyone reading this epic is ignorant of, or has taken tablets to forget what we were working on in the wilds of the North Sea, here is some essential background information regarding the 'hardware' needed to produce oil and gas offshore.

Rest assured there will be no mention of esoteric things like economic importance, bold initiatives, financial risks, seismic surveying, new technology, carboniferous wotsisname, or drilling. However, as it is probably the only serious part of the entire book, you should pay attention, as questions may be asked later.

The areas we are interested in are known respectively, as the Central North Sea, which is roughly opposite the North East of Scotland, and the Northern North Sea, which surprisingly, is north of the Shetland Isles, and a lot higher up than Watford.

Clever people discovered oil in what are known as reservoirs deep under the seabed and so we had to devise a means by which we could recover as much of this precious commodity as possible. This would enable a man called 'The Chancellor' (not Chancer—although I don't know) to sell the oil and spend the money on useful things like politicians' pensions and Jaguars.

In order to get this oil for him, we adopted some of the technology that had already been developed for the Southern North Sea gas fields.

The snag is that the North Sea gets deeper towards the north and the weather becomes more fearsome, especially in the winter, which on average lasts for about nine months of the year. We also found that what the Weathermen, in their search for understatement, called the 'hundred year storm,' actually occurred about once a month. Therefore, in order to combat nature's perversions we had to build things called 'Production Platforms,' which were not only bigger and better than their equivalents in the South, but cost a lot more too.

The other interesting thing is that Nature decided it would be much too simple to create oil and gas close to the seashore, where we could pop across for an ice cream, play on the fair or lie on a deckchair while the oil gave itself up without a struggle.

So where did Nature put the oil? I'll tell you where, right in the middle of the deepest, coldest, stormiest, rainiest, squalliest and most difficult to get to, mass of water in the UK, that's where. Nice one nature.

This stroke of location genius resulted in lots of travel in helicopters whilst wearing the most awful one-piece rubber suits designed by someone with a grudge against humanity. And don't let anybody kid you that flying for an hour or two in a helicopter is fun; they lurch about, make a terrible noise and the seats are an exact replica of those used in a third world infants' school.

Because of this location challenge, a typical oil production platform has to be self-sufficient in life support. It must have living accommodation, medical facilities, electrical generators, fresh water makers, helicopter fuel, fire-fighting equipment, not to mention the myriad pipes and other things required to process the oil and gas. All this, just so the black gold can be pumped either to the shore or to a remote Buoy for onward transport by a giant tanker.

Before going any further, we need to get one thing quite clear; something absolutely fundamental, which drives production personnel mad. The Media apparently has an incurable urge to call Production Platforms 'Rigs'. Every time there's a news item involving offshore platforms, the they trot out the same totally irrelevant introduction, accompanied by footage of a drilling rig showing two or three unidentifiable lunatics covered in mud and pulling on some pipe.

Numerous attempts have been made over the years to persuade journalists to stop this practise, explaining with simple pictures that a rig is not a platform and that indeed, many of the platforms under discussion have long dispensed with their drilling rigs. This is however, to no avail; journalists are convinced that their idea of dramatic footage adds the necessary degree of mystique to a report regarding such dynamic topics as migrating birds or petrol prices. So for the record, Production Platforms are not 'Rigs.' Okay?

That dealt with, back to the platforms. The two main types with which I was associated are known as 'Steel Jacket' and 'Concrete Gravity' structures.

Jackets are fabricated (or, as we in the industry say, 'put together') from welded pipe and fixed to the seabed with steel piles, just like giant tent pegs. All the production plant and the accommodation sit on the jacket and access is from a Helideck located at the very top of the structure. There will be a drilling rig at one end, at least in the early stages, and a flare boom. Oil from the wells has to be exported by a pipeline to shore, or to a remote Buoy to which a tanker is moored.

(There will be many adventures concerned with this particular madness later in the book, don't worry.)

On the other hand, Concrete Gravity structures are massive affairs and have storage facilities for a million barrels of oil in their base. Their immense weight plants them firmly on the seabed without the need for piles. They have similar topsides to jacket platforms and generally export oil directly to shore through a sub-sea pipeline.

I am sure you will agree this is all very straightforward so far. However, these structures had to be built in numerous parts in different locations, then brought together, assembled and transported to their destination. A bit like a really expensive Meccano set (or Lego for younger readers.)

To give an example, on one project I was involved in, we had to manage substructure design in Paris, a construction site in Argyll and an inshore hook-up in Norway, with major equipment coming from Amsterdam, Darlington, Newcastle, Leith, Rotterdam, Lowestoft, Middlesbrough and Glasgow. All of these activities had to be planned, scheduled and resourced and were at times, a nightmare of astronomic proportions that had to be pulled together by the highly skilled and dedicated Engineers in charge.

In practise, the fun started almost immediately and classic industrial practises such as lying, cover-ups, sackings, strikes, unauthorised design changes and falsification of documents were soon employed. Being thrust into this mayhem certainly helped to hone your detection and negotiation skills. Interestingly, despite these minor irritations the enterprise was, and still is, a resounding success and a tribute to the many individuals who had the determination to make it all succeed.

Just to give you some idea of the scale of our construction operations, the water depth ranged from 275 to 500 feet, obeyed no known laws and was bloody freezing. The structures themselves were taller than the Post Office Tower when measured from the seabed to the top of the drilling rig. The steel jackets weighed between 4,000 and 12,000 tonnes but were mere lightweight compared to the concrete structures, which came in at about 300,000 tonnes, with another 27,000 tonnes of modules and equipment perched on top.

Each of these monsters had to be accurately placed in its appointed location; steel jackets were towed to the site on a giant barge, before being tipped off by sinking the barge and finally set on the seabed by a crane of astronomical proportions.

The concrete structures were so big that they had to be submerged in a *fjord* while the topsides were floated over the top and fastened in place. The whole enterprise was then de-ballasted until it was partially afloat and towed out to the location by a number of heavy-duty tugs. Once on site, the storage tanks were flooded and the structure finally settled onto the seabed. Just another day at the office really.

When all of the parts were assembled and tested, the installation had to be 'certified' before we could even allow people on board. This was carried out by an independent Certification Body and could be a fraught experience, a bit like trying to get your house extension passed by the Building Inspector only a hundred times worse. Sad to relate, offering the odd bottle of Scotch did no good either.

In fact, by the time all the excitement died down, many members of the Project Team were in danger of being certified themselves.

It's worth remembering that this huge project was successfully concluded without mobile phones, powerful personal computers, Business Notebooks, 'Intel Inside' (whatever the hell that means,) DVD's, Digital cameras, iPAQ pocket PC's, USB Flash drives or Super Glue. As if that wasn't enough, a thing called the 'wow factor' hadn't yet been invented.

What we did have was an endless supply of Lever Arch Files, Telex rolls, staplers, paper clips and Tipp-ex by the gallon. In fact, we used so much Tipp-ex on documents and drawings that it was impossible to fold some of them. They were so stiff that when you attempted to bend them they cracked, the Tipp-ex fell off in a shower of confetti and we had no idea what the latest changes were, as they had now turned to powder on the floor.

Sometimes I wish the prediction by Ken Olsen, President, Chairman and founder of the Digital Equipment Corporation in the 1970's that, 'There is no reason why anyone would want a computer in their homes'—had been true.

Younger readers may find all this difficult to believe, but you have to remember that when these activities took place, Abba was the sexiest group we had ever seen and no one had ever heard of Kylie Minogue.

Perfect, really.

5

Who's Who

Because these recollections involve scenes and people with names and titles that may be unfamiliar to most of the sane populace, a brief explanation of the more common terms used in the oil and gas industry may be useful, especially for those sad people who go in for pub quizzes.

Veracity has been my watchword in order to ensure the lay reader has an accurate insight into the surreal world of the Offshore Engineer. Don't be fooled by cheaper versions of the truth, what follows is 'gospel'.

Offshore: is an all-embracing term meaning activities that take place 'all at sea.' This also accurately describes the state of mind of the people in the following groups, who were 'offshore' most of the time.

A **Project:** is a close knit team formed by the Oil Company when, having decided that there was money to be made in operating a Production Platform, it gathered together a group of people to build, manage and control the enterprise. Or not, as the case may be.

A Project was generally led by a *Project Manager* who, having been on a number of man-management courses, displayed all the team building attributes of Idi Amin. (Remember him? He was the self-appointed King of Scotland.)

Design Engineers: Their role in life was to understand what it is that the lesser mortals in 'production' needed the plant and equipment for. They then ignored this entirely and produced lots of complicated drawings, so, they believed, that construction people could build the plant to everyone's joy and satisfaction.

This belief, however, was a surreal and delusional world inhabited only by the design fraternity. In reality they tended to produce designs that showed a marked inability to understand customer requirements. This was due to a number of closely held but erroneous beliefs, for example that they were duty bound to 'improve' on previous designs; this led them to have an overwhelming desire to 'innovate' in the belief that they certainly knew better than mere 'valve minders.'

In the eyes of *Production*, these designs were like *haute couture*; nice to look at, but not something one would want to be seen wearing in public. Design engineers were forever doing 'sums' to prove they were right and were therefore not trusted by either production or construction personnel.

It was easy to identify the most experienced Design Engineers, as they all left the Project just before the start of construction and forgot to leave a forwarding address.

Construction: a group of people charged with translating the design into reality. This is the period where lots of welding, hammering and banging takes place. Contractors are normally employed to carry out construction and it is obvious from the start that their objectives bear little or no resemblance to those previously agreed in the contract.

This, by the way, was often despite the fact that the Client had hired a firm of Consultants at huge cost to devise and run team-building exercises. These courses were sub-contracted to some madman whose claim to fame was that he had survived a plane crash in the Andes and who believed that industry would be far more competitive if only everyone had suffered in the same way. At the outset his victims would be reasonably friendly, but after having 'bonded' by playing golf, building a bridge out of used chopsticks and surviving three days under a wet sheet in the Cairngorms, they hated the sight of one another.

Interestingly it was not unusual to find that on the first day of work all the senior members of the Contractor's team had been whisked away to projects on the other side of the world.

Contractors knew this clandestine re-deployment of key personnel as 'asset management.' As far as Construction was concerned, it meant that they had been screwed.

The realisation that the Project team would now be working with a bunch of total strangers who had missed the outward-bound courses, guaranteed that exciting adversarial relationship so beloved of industry.

Commissioning: a group of engineers and technicians charged with the unenviable task of making all the bits and pieces work, such as making the lights come on, the pumps pump and the engines make a noise. Their job is similar to plugging in a new washing machine, only bigger and takes longer. There would be hours of interesting meetings with the Main Contractor and the Project Manager, at which they would have to explain why all the things they have built can't be energised. Generally, this was because (a) they had forgotten to provide drawings that remotely resembled the finished article; and (b) all the pipes and vessels on board were full of discarded safety helmets, welding rods and torn overalls.

Commissioning's strange compulsion to tell the truth about other people's shortcomings often led to confrontation, name calling, accusations of incompetence and threats to the person. Commissioning people are masochistic by nature and are generally looked upon with suspicion by all other parties. You can always tell a Commissioning Engineer, they are the people convinced that 'Star Trek' was a documentary series.

Production: a deluded group of people who believe the platform has been built for them and therefore that they are Very Important Indeed. From the outset they complain bitterly to anyone foolish enough to listen that the newly built installation is not what they asked for. They will moan about unimportant issues such as out of reach valves, doors that won't close properly and having to run halfway across the platform to turn on the generators. They remain convinced that had anyone asked their opinion during design and construction the end result would have been wonderful, while ignoring the fact that their idea of heaven would have cost another £20 million and delayed the project by about five years.

Most Project Managers issue strict instructions to ignore all pleas for changes and improvements by production staff. (Obviously this could be denied later.) In reality, production personnel have little in common with engineers, being mainly concerned with esoteric things like barrels per day, injection rates, the likelihood of fog on crew-change days and annual leave.

Maintenance: a group of people with an uncontrollable urge to tinker with production equipment, believing, erroneously, that they can make it work better. The result is that their endeavours are largely unappreciated and generally misunderstood.

This leads to an uneasy relationship between maintenance and production; the first are convinced that production are constantly trying to wreck perfectly good equipment, whilst the second are equally convinced that the first are a shower of interfering bastards and an unnecessary evil.

Throughout industrial history, there have been many failed attempts to do away with maintenance—meanwhile, production personnel constantly pray for just such a miracle.

Drilling: a group of people who live in a strange twilight world of their own. In general, Drilling ignore everyone on the platform and are, in turn, looked on with suspicion by everyone else. They are led by a *Toolpusher*, who traditionally behaves like a minor despot, as shown by his cavalier disregard for sound accounting practices and his demeanour, which ensured his staff lived in a permanent state of fear.

The old-style Toolpusher's favourite victim was the *Petroleum Engineer*. These were young graduates sent out to provide particular expert advice and do complex sums to prove that the well was performing properly. In performing this role, they suffered from two main problems, one that they were very young and two, that the Toolpusher, believing he was God, hated to be told something he didn't know by a whippersnapper.

In the eighties, an attempt was made to re-brand Drilling by calling the Toolpusher the 'Drilling Engineer' in the hope that this would turn hairy-chested Toolpushers into caring father-figures. The only snag with this strategy was that most of the incumbents were pathologically incapable of embracing such a radical change of personality.

Drilling is famous for failing to grasp the fact that a Platform is not just a bigger than average drilling rig and are only really happy when everybody and everything is covered in mud.

Offshore Installation Manager: always known as the *OIM*, this person is in overall charge of an Offshore Installation. The post is a legal requirement and a number of OIM's were formerly merchant fleet officers, many of whom subscribed to the Captain Bligh style of man management. Others were long serving members of Company staff, generally from the drilling world; luckily, the OIM who made a lasting impression on me was of the old, 'seen it all before' school of management, but more of him later.

Onshore: a group of people looked on with deep suspicion by 'offshore'. They fondly believe they are the decision makers and without their constant support, advice and arbitrary programme changes, 'offshore' would simply crumble into chaos. Onshore staff have a habit of visiting offshore for important meetings on mid week crew change days; this irritates the hell out of offshore staff.

Human Resources: formerly known as the 'Personnel Department' until they could no longer stand being called the 'Anti-Personnel Department,' an accurate if embarrassing truism, by all who fell victim to them.

This re-branding exercise worked until it became obvious that the *only* thing that had changed was the name. Human Resources, or H.R. liked to think that they were the human face of management and that their main aim in life was to help. Occupying a position of importance in an organisation that often seemed in inverse proportion to their capabilities, they clung tenuously to the belief that they were there to bring harmony and dispense largesse to ungrateful employees. Unfortunately, they were in charge of a number of highly emotive issues such as recruitment, contracts, salaries, appraisals and outsourcing.

Most people only came into contact with H. R. when negotiating what was known as their *Package*. This was discussed during recruitment; but the promises and agreements made then often bore no resemblance to reality. When offered, the Package was gift wrapped and extremely attractive with the emphasis on salary, overtime, expenses and holidays. On acceptance of the terms and conditions, however, the emphasis somehow changed to redundancy, downsizing, economies, appraisals and transfers.

Support Services: as far as offshore workers are concerned this means *Catering,* which was managed by an extremely powerful person known universally as the *Camp Boss* (for the uninitiated, the word 'camp' in this context is a noun, not an adjective.) It was his duty to ensure that key issues such as food, towels, laundry, films and, most important of all, duty-free smokes and perfumes were readily available. He lived in a place known as the *Bond,* which was usually under the stairs in the accommodation and had security features to rival those of a Swiss Bank.

Most people only saw the Camp Boss at certain times during the day through a small hatch and consequently had no idea what he looked like in full view.

Probably the second most important support service was known as *Logistics.* This is a term coined by consultants that most people don't understand, until it is explained that Logistics provide hotels, helicopters and taxis.

Effing and Blinding: You may be surprised to discover that swearing, or as the television would have it, 'strong language,' was used offshore. Generally the users fell into two categories; the first being those for whom swearing was a way of life, forming as it did an integral part of their 'normal' conversation.

The best example I recall of this was when working alongside a coal miner who had raised the use of such Anglo-Saxon words to an art form. In fact, he used them so frequently that he had to split syllables in order to fit in his choice of words.

This was okay, but what made it unique was that he was a keen gardener. To hear him describe horticulture in this way was hilarious. If you asked what he had been doing in the garden, he might say, 'I've just planted me daffy-effing-dils;' or another classic was 'I've just bought some lovely chrysanthy-effing-mums.'

Interestingly, one of his mates told me that he didn't dare swear at home, which was why he was renowned for saying very little when out shopping with his wife.

The second category used expletives judiciously or, in some cases hardly ever. Most of our gang fell into this group. (Well I would say that, wouldn't I?) We used swearing to emphasise a point or more generally, to cope with moments of extreme stress.

For instance, lumpy gravy, late helicopters, missing the black or a newspaper with the crossword already filled in, could raise blood pressure to dangerous levels that could be relieved only by cussing the situation, the perpetrators or life in general.

Incidentally, the word 'bastard' in this context was not used as a slur on someone's parentage but was just another useful way of expressing dismay at the recalcitrance of objects, both human and inanimate, to do what you want.

Occasionally throughout this missive I have used one or two of the milder swear words in order to be reasonably faithful to the conversation taking place. I do not subscribe to the view that people who swear have a poor command of the language. In my experience, the addition of certain such words to a phrase or sentence can really emphasise the message. The only rule for success is that they must be appropriate and used with discretion.

5

In the Beginning.......

'Mister Page?'

'Yes.'

'Fly.'

'What?'

'Fly, get on a plane and come to Aberdeen, we will reimburse you for any expenses.'

This rather stilted conversation took place in 1974, in a public phone box in St Helens, Lancashire. The voice telling me to fly belonged to the Recruitment Manager of a major Oil Company and he spoke slowly as though to a child who hadn't understood a word he was saying and to a point, he was right.

I was employed at the time as a Lecturer in the Mining Engineering Department at the College of Technology and was responding to one of many, until then, unsuccessful job applications. The phone call was to invite me to Aberdeen (a place high up in Scotland) for an interview and I had spent an anxious two days trying to find a train or a bus which would get me to their office at the required time.

After much searching of timetables, I had reluctantly contacted the recruitment chap and poured out a lot of nonsense about being unable to make a morning appointment, as there was no way to get there so early in the day.

You may wonder why I hadn't considered driving to Aberdeen, but you have to remember that the average family saloon in 1974 required hours of preparation for such an epic journey. As a minimum, you had to carry spare fan belts, plugs and points, oil, water, a massive tool kit, wiper blades, gun-gum (for fixing holes in exhaust pipes) and assorted jubilee clips. Sadly, all these essentials left very little room for luggage and membership of the AA wasn't an option, especially on a lecturer's salary.

So, there I was in a phone box being told to fly which, to put it mildly, was something of a shock. You see, there wasn't much call for flying when working three thousand feet underground, so all this aeroplane talk was very new to me.

However, in assuring him that I would be there, I tried to sound as though flying was second nature and the option had simply slipped my mind.

My big adventure started some days later when I boarded a plane at Manchester bound for Aberdeen. However, in those days Aberdeen was a step too far and we had to change planes in a place called Glasgow. On arrival at Aberdeen Airport, which was small and resembled something out of a World War 2 aerodrome, I tried to appear nonchalant, took a taxi to the Office and presented myself for interview.

I was now beginning to realise what a sheltered existence I had lead up until then, as even taxis were not part of my daily routine, let alone visiting two airports in one day.

The interviews, of which there were several, went well and I was offered a job. Here again, I began to realise I was entering a different world when the Manager, who made the fateful decision, shook me by the hand and confessed he had no idea what it was I might be doing. He seemed convinced, however that 'I would come in handy,' due mainly, I think, to my extensive mining engineering experience built up working in an atmosphere where the presence of noxious, flammable gas was a way of life, or not, if you lit a match.

The 'icing on the cake' at the end of the day was when a rather lovely Secretary clutching a handful of ten pound notes approached and asked me what expenses I had incurred in attending the interview.

I was obviously now in a sort of dream world; just three weeks earlier, I had attended a seminar at University College Cardiff and during the journey back had taken afternoon tea on the train. On submitting my expenses for fifty pence (or ten shillings) to the College Principal, he refused to pay on the basis that I had eaten lunch before leaving South Wales and would be having tea when I got home.

Now I was being *asked* if I would like some money by the (very) lovely Secretary, with the full agreement of my new Boss. On recovering my non-existent composure, I itemised my costs, fully expecting to be shown the door. Again, I realised that my learning curve was just beginning, as she cheerfully counted out the money and added more to cover an evening meal, which she felt I would no doubt need on the way home.

This earthbound angel then apologised for not having it ready in advance, saying 'What will you think of us being so disorganised?' By now, all I could think of was how I would love to introduce my newfound friends to my erstwhile Principal.

I now realised that any man who insists on using the initials of four Christian names like 'T.E.A.K.' before his surname, would never lightly part with fifty pence.

A short time later, I received a summons to start in the newly formed Production Operations Group. I was put up in a local hotel and found a temporary office, while they decided what to do with me. Interestingly, there were, at the time, insufficient offices for everyone, so a number of new recruits found themselves ensconced in a nearby hostelry, known locally as the 'Greasy Spoon.' Luckily, a role was identified for me and two weeks later, I found myself on the first of many installations in the North Sea.

Construction on this platform was still in progress and fixing the structure to the seabed progressed throughout the day and night. This required the use of a giant hammer to knock huge tent pegs through the steel work and down into the seabed below. Initially I couldn't see the point; I mean if something weighing forty thousand tons needed pegging to the seabed what was I doing there? This question was answered when I experienced our first real storm and the whole structure shuddered like a whiskey drinker being offered dry ginger in his dram.

My first job was to set up a planned maintenance system and assign tasks to the maintenance team. This sounded simple enough, until I realised that a rather feudal system for allocating staff was in place.

At the start of each day, the Production Supervisor would pick his team from the assembled technicians and I could then utilise anyone unfortunate enough to be left over. As a personnel morale booster the system left a lot to be desired, especially as the so called 'maintenance team' now consisted of people not wanted by Production.

It also brought back painful memories of school sports, where football teams were picked from the assembled urchins and, as the numbers waiting to be picked diminished, the realisation dawned that neither of the opposing Captains considered me to be a budding Bobby Charlton.

On about the third day, I could no longer stand the idea of having to work with a bunch of technicians destined to go through life as rejects, so I had a meeting with the OIM and the Production Supervisor. The gist of my argument was that we were paying scant attention to choosing people based on aptitude. At least I think that's what I said.

Whatever it was, I'm glad to say my arguments for dropping this rather arbitrary arrangement prevailed and we all agreed that the formation of dedicated teams would be of greater benefit all round. I was 'fair chuffed' as we used to say in the mines and had achieved my first taste of successful negotiation with someone who was not only an experienced Supervisor, but was also from the USA.

You see, we didn't have any Yanks in the mining industry—displaced Latvians, Poles, Ukrainians and Czechs—yes; Americans, no.

The platform had been designed to cater for steady production, where obviously nothing ever went wrong; as a result there had been a small but important oversight.

The Maintenance Supervisor had no office.

The Management had nurtured a dream, which said they would only need about forty people on board at any one time. However, in reality this figure was exceeded by a factor of three, if you included construction, drilling, commissioning, helicopter crew and assorted specialists. This enforced togetherness made getting food something akin to trying to see a Madonna concert without a ticket. Having been employed to use my initiative, I decided to convert my bedroom into an office, but as I shared this space with Bill the American Production Supervisor, you can see how important it became to hone my negotiation skills.

As we were both 'on call' throughout the day and night and there were just the two of us in the standard four man cabin, I removed the two bunk beds on one side, purloined a dining room table for my desk and mounted two scaffolding planks on the wall above to represent shelving. Remember, this was all before the invention of an Ikea.

Sadly, the discovery that he had lost a dining table almost caused Robbie, the Camp Boss, to resign. However, a promise from Sid, our OIM, that he would order a larger version in oak coloured Fablon as a replacement, seemed to cheer him up, so after a while I was allowed back into the galley and he started to speak to me again.

7

Offshore Life

All self-respecting chaps have a retreat where important things can be saved from their spouse, the charity shop or the refuse collector. These are important collections that require due care and attention; for example, assorted nails and screws have to be segregated, stored and labelled in odd tins. Largely unused tools need to be located on specially designed hangers (or nails, as we engineers call them.) With careful planning and subterfuge, a lucky few have achieved the ultimate in comfort by adding a bench, a radio, a kettle and an old armchair.

Being on an offshore installation is a bit like having your own giant shed, as it has all of the above items on a grand scale. In addition to the standard goodies, your bed is made, your shower is cleaned, massive meals are prepared and with good timing you can eat four dinners every day. What's more, cakes are made for tea breaks and there are films. Not that everything was perfect; sometimes newspapers didn't arrive on time and there might be lumps in the custard.

As well as this you got to play with real man-sized plant and equipment, there were giant cranes to be driven, pumps to be started and stopped, helicopters to fly, tankers to moor, radios to operate and lifeboats to sail.

Visits to Disney World pale into insignificance when compared to the average day offshore; not only that, but you didn't have the additional embarrassment of having to continually shake hands with some idiot in a mouse outfit telling you to 'Have a nice day.'

Continuous cover was provided by two teams named, in a stroke of identification genius, A shift and B shift. Each worked twelve-hour shifts for either one or two weeks alternately. A major benefit of this system was that as each shift took over, they were able to blame 'the other shift' for sensitive issues such as work rate, hand-over notes, competence, crappy films, newspapers, shortage of biscuits and failure to fill the ice-cream machine.

It came as something of a shock for each shift to discover that the other was habitually reciprocating the blame. The only comfort was the belief that the 'other' shift consisted of sub-humans who didn't know what they were talking about. I think it's called mutual respect.

Later on, Management organised a series of 'team building' exercises in an effort to eradicate this problem; and to a degree it worked a treat. On meeting the consultants who were running the course, both shifts discovered a common enemy.

This enabled a truce to be called and both teams were then able to give their undivided attention to the inadequacies of the consultants, before returning refreshed to the fine art of mutual character assassination.

As there were a great many bodies on board during the final stages of completion, I did what all mature males do at such times—I formed a gang. The main members of this exclusive club were Liam and Bert who were both Mechanical Technicians. We were ably supported by Sid, the OIM, Andy the Medic, Ken the Radio Operator, Allan the on-site Helicopter Pilot and George, the Electrical Technician.

Liam and George had some offshore experience, as they were former southern Gas Platform Technicians. However these platforms were small by comparison, being located in shallow water and operating in steady state conditions. Still, compared to the rest of us they were looked on initially in awe, which wore off when we realised they were generally incompetent, just like everyone else.

8

Offshore Democracy

The OIM with whom I first worked deserves a special introduction. Sid was probably then in his late fifties and was a Toolpusher by profession, but I guess nobody's perfect. His appointment as our first OIM was the culmination of a long and varied career, which had been anything but average. He had extensive experience of drilling in many countries, notably the Middle and Far East and just before the outbreak of hostilities, had been involved in sabotaging oil wells in Sarawak in order that the Japanese (who were ignoring Hitler and trying to take over the world for themselves,) couldn't commandeer our oil wells.

Later he was involved in Special Service activities in Europe during the Second World War, as a result of which he was taken prisoner and credited with numerous failed attempts to escape from various German prison camps.

After the war ended, he worked for a time in Australia and loved to tell the story about his failure to drill a shallow water well for a farmer in the outback, despite having all the latest drilling facilities.

Apparently they were doing some exploration drilling on the farmer's land and each day he would come along to watch the proceedings. Thinking he would take advantage of all this expertise and technology, the farmer asked if Sid's crew would drill him a water well before they left.

Sid apparently then had a brainstorm and told the farmer they would be pleased to do so, a promise he began to regret almost immediately.

It transpired that no matter where they tried, water resolutely failed to materialise and the more they drilled, the more the farmer became critical at their lack of success. As the days went by Sid's lack of success became a serious embarrassment both to him and the Company. Each time he pleaded with the main office to be released from the task, they were equally adamant that of course we could find a simple thing like water for a farmer and so they insisted on him continuing.

It was, as Sid ruefully put it, 'A matter of credibility. Imagine being remembered throughout Australia as an Oil Company who couldn't find water.'

I asked Sid if he ever did find any and he said that one morning having noticed a small muddy puddle around the drill pipe, he gave the order to pack up and get the hell out of there. One of his operators asked if he should notify the farmer and Sid replied 'Not bloody likely, we'll telephone him and the Company when we get to the nearest town.'

I was obviously going to be in the hands of a master craftsman.

Sid's unflappable approach to managing a disparate group of workers during construction, commissioning, drilling and plant operation, whilst wearing carpet slippers, was a revelation to me.

Some of my fondest memories of Sid centre on his interpretation of an edict from 'Onshore' that he should hold regular Staff Committee Meetings. Initially, he was reluctant to comply and muttered all week about interference and bureaucracy. He told all and sundry that if anyone had a problem, they only had to tell him and he would do his best to solve it. This latter statement turned out to be a load of rubbish, but we were still very naive and easily impressed at the time.

However, being an 'old hand' and recognising the futility of protest, he inaugurated a series of meetings and 'volunteered' me to be his Secretary.

The Committee consisted of representatives from Production, Drilling and Catering. Sid's idea was to pick the most amenable people to be members but, as we shall see, either he was rubbish at picking the right people or they became power mad on being elevated to this new status.

However, due in no small part to Sid's ability as a Chairman, the meetings were very enjoyable, except when they landed me with a number of challenges regarding appropriate (but not necessarily truthful) wording of the minutes.

My job was to translate Sid's tendency to treat each issue as though it was singularly unimportant, into crucial sounding text that vaguely identified actions to be taken.

This required me to use what is known as 'creative' wording and, as Sid was generally the only person who had the authority to provide solutions, I had to be doubly careful not to make him the 'action party' for everything.

Just to make things even more interesting, the minutes also had to convince Onshore of our awesome grasp of critical issues.

Unfortunately for me, Sid tended to display a cavalier attitude to any need for urgent action. Juggling with the facts proved to be something of a challenge and required me to tread a fine line between flippancy (not treating things with due reverence) and sycophancy (treating Sid as our saviour).

I also quickly realised that, whenever he was challenged about an item in the minutes, Sid had no compunction in asserting, rather grandly, that the Secretary must have failed to reflect what he had actually said. This was *despite* the fact that I made sure he approved the draft minutes prior to issue. Obviously I still had a lot to learn!

Although most of the committee quickly realised that Sid's promises were not always translated into tangible improvements, the Catering staff were the exception.

They looked upon Sid as a sort of God and it took them longer than most to come to terms with the fact that although, in their eyes, he was the 'fount of all knowledge,' their problems remained largely unresolved.

Sid of course, had realised that our 'committee' problems paled into insignificance when compared to the issues concerning Senior Management onshore. I suppose being faced with rapid expansion, Tony Benn, major construction costs, suspicious Partners and massive logistics problems, made it relatively easy for them to ignore our cries of woe.

Sid believed the initiation of a staff committee was simply a means of obtaining information about what was bugging us, in the naive belief they could avoid making the same mistakes on the next generation of platforms.

In hindsight this was obviously wrong; the ability of the average worker to dream up new and interesting injustices is endless.

We, on the other hand, being new to this expanding enterprise, believed that our platform represented the centre of the known universe. The cruel reality took some time to sink in, but in the meantime we felt quite justified in criticising the lack of any immediate and sympathetic responses from 'that lot onshore'.

Obviously, as we were being conned by a master, this critical attitude didn't apply to Sid.

There were a number of crucially important grievances that, in the eyes of the staff, required urgent action. Each was debated with deeply held opinions and it will come as no surprise to find that the longest running unresolved issues were films, newspapers, a lack of soft drinks, peanuts, popcorn, ice cream and the finishing time for supper.

Of less importance were a number of unrelated problems such as washing machines, ventilation, erratic shower water and mattresses.

As the following examples show, we had a sound grasp of priorities and a commendable willingness to devote time and resources to significant problems. Already we were displaying a fine example of *esprit de corps* and democracy at work.

At eight o'clock each evening the Recreation Room was transformed into a Cinema. This involved much switching around of seats, commandeering chairs from nearby offices and Andy the Medic carting the projector, re-winder and film from the Sickbay. The screen was unfolded and hung on the wall, the first reel was mounted on the projector and we were ready for the big event.

With a totally unfounded air of anticipation, everyone who was off duty scrambled for a seat, sorted out the oranges, apples or sweets to be consumed and waited impatiently for the Andy's order to 'Switch out the lights.'

It was just after this that the inevitable sense of disappointment and anti-climax started. Generally the film was of the 'Where the hell did they get that from' variety and starred, if that's the right word, people of whom nobody had ever heard.

This aroused considerable passion at subsequent committee meetings and centred on the inability of the Personnel Department to pick films that had been made within the living memory of anyone in the audience.

The seriousness of the situation was amply illustrated during one meeting by the maintenance representative, who had volunteered to set up an exchange system with a Crane Barge, which at the time was moored alongside.

With growing emotion he recounted the story of how, initially, his idea had been well received by the Admin guy on the Barge and an informal agreement was struck whereby we would swap them film for film, thereby doubling our choice. Brilliant.

As a result of this highly commendable initiative, we were now able to watch a selection of brand new films, such as Jaws—very appropriate for the middle of the North Sea.

However, he said, on his latest visit and clutching three reels of our finest, the Admin guy had examined the title and told our man to 'Sod off.' He had explained that he wasn't prepared to face a possible lynching by his audience, which apparently was very likely, should he have the nerve to show any more of our 'designed to kill the cinema' epics.

As realisation dawned that our access to proper films was short-lived, a heated debate ensued, during the course of which issues were raised regarding the incompetence of our negotiator, the dubious parentage of the Barge crew and a minority philosophical view that this was simply 'our destiny.'

This last opinion was not debated to any great extent and on Sid's instructions, was not minuted, probably due to the fact that the comment was aired by the catering representative who, it was suspected, had a penchant for smoking strange smelling tobacco.

It was also Sid's opinion that management onshore may be somewhat alarmed to find we had time to debate Maoist principles regarding fatalism.

As with most of the really crucial issues we raised at our meetings, the minutes invariably identified Sid as the 'action party'. He would then promise to communicate our concerns to management, in the hope of a satisfactory resolution. Looking back, I now realise that we never actually saw any of this correspondence and when questioned about progress, Sid would say something vague like 'I've told those bastards onshore that we are not prepared to put up with the situation any longer and something must be done forthwith.'

The use of such robust language to management convinced us that Sid was fighting the good fight on our behalf and failure to have the situation resolved was not his fault.

We clung to this naive belief for many months. It was a fine illustration of both Sid's political acumen and our pathetic faith in his integrity.

Before moving on to an equally contentious issue, it is worth recounting another film saga, which perfectly illustrated Sid's management style.

Allan, our helicopter pilot, arrived one day clutching a heavy canvas bag which he placed on Sid's desk, saying, ' I've just come from refuelling on one of the floating drilling rigs in the area and they have agreed to lend me this.'

The bag was opened and there, wrapped in Allan's underpants, was 'The Godfather.'

Bloody hell, we stared at it in disbelief for some minutes and then Sid took charge.

'Right,' he said, 'Get Andy in here.'

Andy duly arrived, hoping there was a medical challenge to be conquered, and then he too stared in amazement at the reels on the desk.

Sid launched forth. 'We need to plan this carefully. First, you lock this in your drugs cupboard, then we need to have a special showing with printed notices informing everyone and, as it's a long film, we need to start at 7-30, so we have to hold back supper for another half hour'.

Once again, Sid had demonstrated his ability to grasp complex situations and immediately organise a plan of action. It's no wonder we were in awe of this intellectual giant who could, with scarcely a pause, manage the 'big picture'.

The showing was scheduled for Saturday night and by about 7pm the Recreation Room was filling rapidly. Soon all the seats were claimed and there was standing room only at the back. Many people had showered and shaved in recognition of the big event and sweets, apples and oranges were being bartered in anticipation of the treat to come.

As was fitting, Sid and Allan were given seats in pride of place and the Medic, having retrieved the film from the security of the drugs cupboard, was loading the first reel onto the Projector.

Now, our projector was not exactly 'state of the art'. It was prone to blowing its bulb during a performance and Andy's status was in direct proportion to his ability to produce a spare at the crucial time. The bulb also became extremely hot during transmission and would readily set the film on fire should the projector stop in mid reel (a not uncommon occurrence.) This would be advertised by smoke blotting out both the screen and the Medic and required rapid action on the part of a maintenance technician to isolate the smoke detectors, if we weren't to be drenched.

In addition, the projector was also prone to snap the film during transmission, which was a nightmare scenario for Allan, who had guaranteed to return the film in pristine condition. Again, a good Medic was expected to carry both a film-splicing tool and a thermometer with him at all times. Splicing shortened the film with the consequent loss of some important dialogue, but no one seemed to mind as we were more interested in what Sophia Loren was or wasn't wearing—not what she was saying.

Notwithstanding the potential for fire, flood or breakdown, we were at last ready for transmission. Oranges had been peeled, sweets unwrapped and passed around and the lights were dimmed. We settled into the first reel—wonderful, a new film with lots of killing and a great story.

At the end of the first reel, the lights went up and Andy changed to the next reel, whilst there was a buzz of conversation, centred on the fact that no one had understood a word that Marlon Brando had said.

The lights went down and we were off again. The next two reels came and went without mishap, but conversation in the interval took on a puzzled aspect, questions like 'where the hell did he come from' and 'I don't remember that' filled the air.

However, we were soon off and running again, that is until part way through the next reel. There now began a groundswell of discontented mutterings amongst the audience, culminating with a forceful instruction from Sid to 'Stop the bloody film and put the lights on.'

He turned to Andy and echoing our worst fears, said 'You blasted halfwit, you've put reel four on before reel three.' There was a stunned silence, oranges were discarded and the awful realisation dawned; no wonder we couldn't follow the bloody story.

Andy, who at that moment feared for his life, could only stammer, 'It's not my fault, I'm only used to films with three reels and anyway, I was too busy trying to understand what Marlon Brando was saying, not when the sod was saying it.'

There then ensued a rather unstructured debate about the merits of starting the whole film again, but Sid put the blocks on this proposal as he said it would make everyone late for supper.

Thereafter, the only sound in the deserted Recreation Room was of the discarded reels being re-wound by a chastened and somewhat isolated Medic using an antiquated hand winder, fastened onto a piece of wooden packing case. Usually there was no shortage of volunteers to help out with this chore, however as a result of the debacle the room had emptied in record time.

Andy was left to complete this thankless task alone, except for Allan, who had promised to return the film the next day and wasn't about to be compromised. So ended my first attempt to see 'The Godfather' and the harsh reality of offshore conditions began to dawn.

Second only to the seriousness of sub-standard films was the ongoing saga of missing newspapers. Throughout my stay, this topic alternated with films as item one on the committee agenda and was an issue of fundamental importance to people who, when onshore, never read a paper from one week to the next.

Let me explain the system, if that's the word, for delivering papers offshore. Someone (generally a failed trainee) from the Personnel Department was sent to a local paper shop and depending on what was available, bought a random selection of newspapers. They were then wrapped in brown paper and delivered to the post room to be put in the platform's mailbag.

All well and good so far. However if, as was suspected, the person charged with this task was of a lazy disposition, most of the papers would have gone by the time they got to the shop. The more enterprising souls, not wishing to fail in this important task, would come over all entrepreneurial and supplement the missing papers with an eclectic selection of magazines, with gripping and highly relevant titles such as 'Dolls Furniture Collector' or 'The Pig Breeder's Gazette'.

The problem could be compounded if the lazy sod (not my words) missed that day's post, since there may not have been another delivery for three or four days.

I recall that on one occasion, we received a copy of 'Woman's Own,' which surprisingly was left unread for some time in the recreation room. That is, until some pathetically bored soul opened it and came across an 'Agony Aunt' column at the back.

He found one sensitive letter signed only as 'Worried Liverpool,' upon the subject of which a growing number of staff believed they were qualified to offer advice. Covert enthusiasm for the magazine grew to such an extent that one guy, desperate to know whether the agony aunt's advice reflected his own, asked if the magazine could be ordered regularly under a plain brown wrapper.

Sid, ever mindful of the need to preserve our 'offshore tigers' image, said there was no way in which he was about to become a laughing stock in the Office. His alternative suggestion that the guy should get his wife to order it for him at home was not well received.

So there we were, with newspapers arriving late or not at all and when they did they were of the wrong political persuasion or failed to meet the aspirations of the prospective readers with regard to pictures or football results. You may now begin to grasp the seriousness of the situation. How can people be expected to avoid twelve hours of hard work when there is no up to date copy of the Sun awaiting them when their shift is over?

No wonder then, that papers occupied a position second only in importance to films—the situation was clearly intolerable. Could Sid come to our rescue?

Sid had a vested interest, as he loved reading either the Times or the Daily Telegraph; politically he seemed to occupy a position somewhere to the right of Genghis Khan.

Our vision of a happy OIM was to peer into Sid's office and see him with the newspaper open on his desk, a large mug of tea at the side and his favourite granddad slippers on his feet. With glasses perched on the end of his nose, he would devour the news and, if all was well, turn to the crossword.

We learned to be cautious about disturbing him on these occasions. No trivial queries regarding lifeboats or breakdowns could be raised and you had to come up with correct solutions to a particularly obscure crossword clue if you wanted his full attention.

Sid, of course, had the papers delivered to him for distribution and his increasingly frantic search among the tabloids for his beloved broadsheets could be somewhat nerve-racking to watch.

So, in our search for a solution to the newspaper problem, we had a considerable ally in Sid, especially when *his* papers failed to arrive. On those occasions, we had no need to convene a committee meeting; Sid would fling himself into the Radio Room, stop Ken from whatever else he was doing and fire off a vitriolic telex to shore. He would then make copies for all and sundry, both as a demonstration of his tenacity and as a means of calming himself down.

This proliferation of evidence made writing the minutes of future committee meetings much easier, the only danger being that Sid might by then have forgotten his tirade and be in full denial mode.

There is a nice postscript to the ongoing newspaper saga. After I had left the platform for a new assignment, I received a letter one day and inside was a copy of a telex sent to Sid from Onshore, which read:

Urgent: to the OIM
'Apologies all round. Papers were found this morning under some freight at Scottish Express Cargo. Will be out on 1ˢᵗ flight on Monday.'

Underneath, in Sid's well-known scrawl were the words,

'Dear Brian, what do you think about the above – Victory after two years. As a

member of the old team I thought you would like to know. They must be a bunch of sissies in transport now.'

As I said earlier, sometimes you are lucky enough to meet a manager who is in the right place at the right time!

Before we leave newspapers to look at other crucial issues, there is one more story worth recounting. It involved Sid and I taking on the might of the *Times* in defence of our Company. Unfortunately, this exercise in gallantry didn't turn out exactly as we had hoped.

Our platform was a steel jacket construction and therefore had no oil storage facilities. What we did have, some six miles away, was a thing called a 'Remote Buoy,' floating like a giant cork and anchored to the seabed by huge chains. This was connected to the platform by a pipeline and when a ship attached itself to the Buoy we could pump oil across and fill the tanker. To get onto the Buoy we had a small helicopter on permanent loan. It was manufactured by Messerschmitt-Bolkow, a German company (how did I guess?) and was a superb, state of the art machine with a full plexiglass dome at the front, one seat at the side of the pilot and room for two or three passengers in the rear.

In flight it was extremely fast and manoeuvrable and we had lots of fun in it. Allan the pilot and Colin his mechanic lived on board with us and crucially, when not in use, the Bolkow was parked on one side of the helideck. This arrangement prevailed without mishap for some considerable time until that is, the *Sunday Times* decided to go into investigative mode.

Sid and I were both avid readers of the *Sunday Times* and held it in high esteem, particularly in the days before it became part of the Australian empire. Imagine our disbelief therefore, to read one Sunday an *Insight* article that implied that our Bolkow was a hazard to incoming passenger helicopters.

Now, at that time, the *Insight* Team had an excellent reputation for uncovering dastardly and unsavoury events, perhaps the most notable being their expose of the Thalidomide drug scandal. So to read that we were also considered guilty of placing lives in jeopardy was something of a shock.

The article was entitled *Hazards of an Oil Rig Run* (again with the 'Rig'). Their contention was that a pilot had refused to land on our platform due to the intensity of the flare and the fact that our Bolkow was parked in the centre of the helideck and could not be moved. The banner headline was accompanied by a photograph of the Bolkow sitting stationary on the deck.

Sid and I festered all week and on arrival met in the corridor, both of us waving the article at each other and full of righteous indignation at what we considered to be biased, ill informed and inaccurate reporting.

It was obvious that immediate action was required to rectify the situation and re-establish the good name of our platform.

'Right,' said Sid, 'First things first, get the lads to work, order up some tea and let's write a strong letter to the Editor apprising him of their calumny.' (What he actually said was 'Let's teach these bastards a lesson,' but you get the drift.)

Our essential point was to refute their contention that our Bolkow was usually parked in the centre of the helideck. Furthermore, on the day in question, the wind was light and the flare was blowing away from the deck, and so on.

We waxed lyrical about our concern for everyone's safety being paramount; how radio beacons were operated to aid navigation, advance information regarding current status was telexed to the helicopter base, the OIM would not countenance hazardous approaches and that we were competent in the management of all flying matters.

It took us hours to compose, refine and finally draft. We re-read it out loud and smiled contentedly at our combined genius with words. One thing to bear in mind was that the letter, which consisted of some three close packed pages, was hand-written, as we had no access to a typewriter. Besides, we had a pathetic belief that maintaining the utmost secrecy was of paramount importance when dealing with the Press.

It was agreed we would jointly sign the letter and that I would post it on my return onshore. All that remained now was to buy the *Times* next week and enjoy their grovelling apology for a callous (and careless) misrepresentation of the facts, compounded by their obvious failure to consult with us prior to publication.

Strangely, as far as we were concerned, there was no mention of our well-reasoned protest in the next edition, but we did receive a jointly addressed letter from the *Insight* Editor, thanking us for our comments and expressing his regret that we took exception to his report. He ended by saying that they had consulted widely prior to publication and felt that they had reasonably reflected the situation.

Obviously, Sid and I didn't agree.

Alarm bells should have rung on a more careful reading of this last paragraph, but such is the heady influence of putting the Press in its place, they didn't.

During that week we received a Telex saying that the Head of Public Affairs wished to see us on our return ashore. Sid said we would probably be commended for defending the good name of the platform and our courage in taking issue with a major newspaper.

Having arrived onshore, we combed our hair, zipped up our jackets and knocked on the door of his office, a palatial affair on the top floor. His secretary showed us into his room, which seemed to consist mainly of dark oak and polished leather. The walls were covered in pictures of our offshore installations, both in place and under construction and predominant among the collection were a number of large photographs of our host shaking hands with famous people.

It was all somewhat different to our office walls, Sid's being tastefully adorned with dog-eared pictures of Arnold Palmer torn from an old calendar and mine covered in second hand scaffolding planks.

His Secretary, on ushering us into the great man's presence, smiled and asked, 'Would you care for a coffee?' But before we had time to graciously accept, our man stood up and growled, 'No they bloody wouldn't, just leave us and shut the door.'

He then removed his suit jacket, which had the largest pin-stripes I had ever seen and without any preamble about the weather or football, went mad.

'What the hell do you two idiots think you're playing at?'

'Whoever gave you the right or authority to speak to national newspapers on behalf of the Company?'

'Why don't you just mind your own business and concentrate on drilling for oil or whatever the hell you are supposed to be doing?'

'I don't appreciate getting critical letters from the *Sunday Times* just because some interfering fools have strayed into areas that don't concern them.'

The tirade continued for some minutes until he fell back red faced and exhausted.

As we couldn't get a word in edgeways we just stared as, still twitching with emotion, the pin stripes continued to rotate violently around his legs.

He then drew breath and went even further down in our estimation by stating that he and his PR advisors had reviewed the draft article prior to publication and had pronounced themselves satisfied with the contents.

Still virtually speechless with both rage and frustration at this apparently spineless approach to handling newspapers, Sid, ever the diplomat, tried to regain the initiative by stating that nevertheless we thought the article to be 'A load of inaccurate rubbish.'

The Head of Public Affairs stared at us both, drew in a deep breath to regain some semblance of control and said, 'Listen, you two idiots, haven't you heard a word I've said? Let's make a deal shall we? I won't tell you how to produce oil and you won't tell me how to deal with the media. So, in future if you want to write to the papers, stick to politics, it's safer for all concerned.'

'Now, have I made the bloody position clear enough for you'?

Sid and I looked at one another, and lost for words at such ingratitude, muttered 'Yes' and left with as much dignity as possible. We even forgot to mention our expenses for staying overnight in a hotel, although this was something Sid was quick to rectify on our return offshore. Some things are too important to be ignored.

The episode convinced us that C. P. Snow was right when he postulated the theory that there were two cultures, science and the arts, each one of who failed to understand the other.

However, Sid being Sid, on return to the platform for our next trip and still nursing his wounds sent a telex to the Transport Department, which contained the somewhat telling paragraph:

'......*In view of my previous correspondence and this one, it seems reasonable that they should not send out a pilot who is unprepared in any circumstances to land alongside the Bolkow and certainly if unprepared to do this, they should make sure of the position before taking off.*'

Game, set and match to Sid and dignity regained!

Interestingly some weeks later, our Public Affairs man, showing an admirable ability to forgive (or needing a favour,) asked us to show a group of newspapermen around our platform.

The visit was part of the Company's strategy for informing the public about the new drive for oil and gas in UK waters. I immediately thought 'Oh no, Sid will wreak a terrible revenge and we'll both be sacked.' However we agreed (well, Sid did) and a helicopter was chartered for about seven reporters from numerous regional papers, including *The Newcastle Journal*, *The Manchester Evening News* and *The Liverpool Business Post*.

To my relief, the trip went well and afterwards Sid received a letter from our man thanking us for our help and hospitality.

Much to my surprise, he also contacted me and asked if there was a memento I would like for my help in guiding the visitors a round the plant. Now when I had been in his office I had noticed a large aerial photograph of our platform on his wall. Chancing my arm, I said I would like a copy of that particular picture to show the folks back home.

'I'll see what I can do,' he replied.

Some time later I arrived home to find he had delivered the actual framed picture from his office and despite several house moves it still hangs proudly on my wall. Who said big oil and gas men aren't sentimental?

I heard later that Sid was asked the same question and apparently a bottle of twelve-year-old malt changed hands. Obviously Sid was also a sentimental old fool but with more immediate requirements.

A constant item on the Committee agenda was the level of complaints about the washing machines. These units had been installed primarily for the drilling crew who seemed happiest when covered in mud, but the other major user was Robbie, who needed them to wash towels, sheets and the cooks' overalls.

They were operated by a member of the catering staff who suffered from the delusion that the secret of a successful wash was directly proportional to the amount of soap powder, bleach and heat the machine was capable of achieving.

The end result was both obvious and disappointing; the drillers' overalls suffered a major reduction in size and changed from deep red to a nice shade of pink. This had a knock-on effect on the towels and sheets, which came out a sort of tie-dyed ochre, as well as being shrunk. None of this was helped by the operator's determination to fill each machine to capacity regardless of colour, use, or dirt content.

You have to remember that our catering staff generally carried out reasonably menial tasks such as making beds and cleaning floors and were not selected solely on the basis of technical know-how. By and large, they were conscientious and cheerful, but detailed operating instructions were not their favourite bedtime reading.

Having run out of towels and just quelled a near riot in the Drilling Department, we gently challenged our latest laundryman about his failure to operate the plant successfully.

His reply was short and punctuated with numerous references to the almighty, the gist of which was that he had never used a sodding washing machine in his life and as sure as hell hadn't come offshore to learn.

Sadly, it seems we had failed to meet his aspirations for North Sea tigerdom and at the end of the trip, we never saw him again.

Fearing that he would soon have no staff left, Robbie made a plea for the Company to replace the sophisticated domestic machines with simple, robust models.

At this point Sid, who had only been half listening, said, 'I remember when I was abroad we had a giant washing machine. It was made of stainless steel, with just one programme and was powered by a huge electric motor which never broke down.'

Robbie couldn't believe his luck. 'That's just what we want,' he said, 'Where can we get one?'

'I've no idea,' replied Sid, nonplussed 'Oh I remember now, it was about twelve years ago when I was in Borneo.'

Realising that he had just demolished Robbie's hopes for a washday heaven, Sid made things even worse by promising to bring out the Hotpoint Technician to have another go at fixing the problem.

As we were all relatively new to the offshore scene, the Company had seconded a couple of experienced Americans to help us settle in and show us how to boast properly

We thought it would be good for transatlantic relations for us to invite them to a meeting, during the course of which one of them raised the issue of 'Peanuts.'

Apparently, on American platforms, peanuts, popcorn and ice-making machines were standard issue, without which there would be tantrums, despondency and appeals to the President (of the country, not the company).

The omission of such morale boosting essentials obviously made us Brits feel somewhat inadequate. Although we were used to the stereotype of America where anything we had in the UK, like buildings, lakes or crime, was bigger and better over there, not to have staple commodities such as peanuts was somewhat embarrassing.

We began to come up with valid reasons as to why we didn't have the items in question. Peanuts were only used to feed birds and popcorn was largely unknown to us. Remember this was 1975 and people didn't eat food out of big plastic bins in the Cinema. In fact, boiled sweets were the treat of choice, having first removed the wrapper while the sweet was still in your pocket. The one exception might be a small box of Black Magic if things looked promising on the back row. As for ice making, most of us felt that worrying about the availability of ice in the middle of the North Sea was fairly low on our list of priorities.

Our American friend listened to this rubbish for a while, shrugged and said, in that annoying way of superior beings 'They told me you Limeys are still in the dark ages, I just hadn't realised just how Goddamn bad it would be.'

At this point, Sid stepped in, probably realising that a heated discussion on the role of the USA in Vietnam was about to take place and promised to 'See what he could do.' Once again, I had to be creative with the minutes.

However, just as we were bringing the meeting to a close, Robbie startled everyone by saying that he once had about twenty boxes of peanuts in his store, but as no one had ever asked him for any, he had been forced to eat as much as he could and had fed the rest to the seagulls.

At this point, Sid lost control of the meeting and I lost control of the minutes.

9

Over the Hills and Far Away

One day at home the phone rang, as it does when you pay your bills, and my Boss said, 'We want you to go to America and take a look at some new Gas Compressors.'

This to a chap who until recently had thought a long distance trip meant going to Bristol in one day. My Boss's idea was that six of us from two platforms would spend a week or so at the Compressor Factory and learn to love the machines prior to us squashing lots of gas with them and sending it back into the reservoir (don't ask.)

We arrived in the Land of the Free without any trouble, other than the fact that my luggage didn't. The compressor factory was located in a town called Olean in New York State and we were accommodated in a splendid hotel close to a golf course. Each morning, after the biggest breakfast we had ever seen, a huge car would pick us up and take us to the works.

Coming as we did from northern climes, humidity was just something we had seen in jungle films where it was part of the 'white mans burden,' punctuated by Hollywood phrases like 'I'm going mad I tell you, it's the damned heat, Carruthers.' Stepping out of our air-conditioned hotel however, we began to empathise with the characters in the films, as suddenly our nice crisp white shirts became sodden dishcloths and our glasses steamed up.

Despite this minor setback we were well looked after and had time to take in some important cultural sites such as bars and restaurants. There were chauffeur driven limousines at our disposal and Jake, our driver, turned out to be a retired 'Cop' with a fund of stories about his time during liquor prohibition in the States.

Apparently New York State had been a direct route for the 'bootleg' lorries from Canada where the 'hooch' was made, down to New York where the stuff was distributed.

You may have noticed my easy use of words such as 'bootleg' and 'hooch'. This is due to a thorough grounding in early gangster films, starring Edward G Robinson and James Cagney. My easy familiarity with the vocabulary used by organised crime paid off and enabled me to converse easily with our driver. And people think that regular cinema attendance during my formative years was misspent. Hah, as if.

Jake told me he had been a young Policeman during the prohibition era and the heavily armed convoys would regularly drive through his territory at night.

One of our intellectually challenged chaps then asked what turned out to be a very naive question.

'Didn't you try to stop them?'

'What, and get my Goddamn head blown off, are you crazy? We just did what every other cop did and turned the other way. Man, this is the Mob we're talking about.'

This was followed by a period of silence and as we digested the information, Jake's face turned from puce to pale orange and the veins in his neck slowly stopped pulsating.

I was going to ask him what he thought of Rod Steiger's performance in *Al Capone*, but on seeing the throbbing veins, thought the better of it.

Just before leaving Olean we had dinner in a superb steak restaurant, where the menus were about two feet across and described ways of cooking beef about which dreams were made. When the waiter arrived to take our order he pointed out the 'starter' buffet, which consisted of a row of cold cabinets about 60 feet long, and said we were to help ourselves. He began taking orders for the main course and I asked if I could have a T-bone steak, having only ever seen one before, in a John Wayne film.

The waiter replied, 'Certainly, sir, how would you like it cooked?'

'I'd like it very well done please.'

Again, one of the lads opened his mouth too soon, offering his expert opinion that having a steak very well done was a sin and the meat would be dry and tough.

'Sir,' announced our waiter, staring at our 'expert' while addressing me, 'You sure can have it well done. We will sear the outside to retain all the juices and then cook it to your exact requirements. Don't worry, it won't be dry and it sure as hell won't be tasteless.'

It was at that moment I fell in love with America and possibly, the American dream.

We were in a bar one evening and armed with our new knowledge regarding the role and close proximity of the Mafia, we started to chat to the lady behind the bar.

She mentioned that the bar had been used by truckers for many years; so inevitably we brought up our favourite subject. Subtle questions along the lines of 'Did the Mafia use this bar? How many did they kill? Was it just like the Godfather?'

To our astonishment she made a shushing sound and looked somewhat nervously from left to right. Leaning over the bar she said, 'Please, we don't use that word here, we just refer to them as 'the friends.' You never know who might be listening.'

Realising that she was serious, we started to look around ourselves, uncomfortably aware that we risked ending up in the foundations of some new office block. Watching gangster movies in the cinema was one thing, becoming a part of the story was quite another—so we did the British thing, waved to everyone in sight and quickly left the bar.

On our final day we were taken to see Niagara Falls, a journey of about ninety miles. It was a leisurely drive and en route we stopped at a diner for coffee. The lady serving us recognised our accents and asked where we were going. We said we were going to see the 'Falls.'

'Oh my,' she said, 'I've never been there. I feel so ashamed here we are only 30 miles away and here's you boys coming all this way just to see our sights.' We didn't like to tell her we hadn't come just to see the Falls, but why ruin someone's illusions if you don't have to?

As we were on our way once more, Jake asked us where we would like to eat. The consensus was that as we had overdosed on steak, it would be good to have a real American Burger, with fries and coke (the drink, not the stuff you put up your nose.)

To our amazement, Jake picked up a microphone and went into CB radio speak. He gave a call sign and announced his presence by saying he was on his way to Buffalo with some 'good buddies' and did anyone out there know where to get the best burgers in town.

We were then treated to a stream of calls from truckers all of whom had advice on the best place to eat.

Thanking them for the information, Jake then asked for directions, which were soon forthcoming, and we pulled up outside a huge diner. As it happened, the information provided by our unknown callers was spot-on and we had a superb giant burger with all the trimmings, including coke.

Even so, we later confessed to being somewhat embarrassed by all the 'good buddy' stuff. I suppose it just shows what an up-tight nation we really are.

Jake showed us the awesome Falls and pointed out that one side was in the USA and the other was in Canada. I asked him how you could tell the difference between the two nations.

He replied, 'Shucks, that's easy, a Canadian is sort of like an American, but without a gun.'

My first experience of Americans at work was with a guy called Bill who hailed from Bakersfield, California. He had been seconded to us in the early days to provide much needed oilfield experience and was everyone's image of what an American should look like; he dressed in real Levi jeans, a check shirt made of thick material with pearl buttons, pointy boots with elaborate tooling on the sides and 'shades.' As if that wasn't enough, he had long blond hair and a beard to match.

I invited him to spend a few days with my family and at the end of our shift, we drove down to Cheshire, which I had to confess was smaller than California.

Bill's idea of a 'little runabout' was an Alpha Romeo, again not a motor with which most of us were on intimate terms in those days. I remember my wife losing her *sang-froid* for a moment when, mentally discarding her Hillman Imp as the car of choice, she asked if she could take the Alpha to work at her College.

For the benefit of those ignorant of motoring lore, the Hillman Imp was made with loving care by Glaswegians in Linwood. It boasted a modified aluminium Coventry Climax fire pump engine in the boot, had a remarkable propensity to go on fire and, due to the clever camber setting, suffered from terminal tyre wear after about ten or twelve miles. Another triumph for the British motor industry.

As this was the family's first exposure to a real live American, Bill went down a storm, as they say in the movies (sorry, films) and did wonders for my parental street cred. As far as my daughters were concerned, Clint Eastwood himself had arrived; not only was Bill a living, breathing cowboy, he even spoke like James Stewart. The only problem was that they wanted to take him to school; I think their rationale was there was no point in having a treasure if you can't show it off. It's just as well that grown-ups don't behave like that. My son (the pragmatist) had a simple request; he wanted to be allowed to reload Bill's Colt forty-fives. So much for the influence of television.

Obviously we had to show Bill a good time and so we decided to take him to a Restaurant in Chester and overawe him with our sophistication. There was only one establishment that would fit the bill; it had to be a Berni Inn.

Remember this was 1975 and our idea of haute cuisine had only recently been 'Chicken in a Basket' and 'Prawn cocktail'. This latter delight consisted of a wine glass (sophistication in itself) stuffed with chopped lettuce, about 5 or 6 tiny shrimps (pretending to be prawns) cleverly hidden in the greenery, covered in a glutinous bright pink sauce. We gourmets knew this as a 'Starter.' Oh, and obviously proper thick chips, not 'French fries,' accompanied the chicken.

Subdued lighting was achieved by placing candles in raffia bound Chianti bottles. Strangely, we never found a pub that actually sold Chianti and assumed that the empty bottles had been acquired by raiding the bins of a local Italian restaurant.

The meal would be accompanied by a vintage bottle of Yugoslavian Riesling (at least three weeks old) not chilled; this would be taking pretension too far.

Anyway, by the time Bill arrived, we had moved on from this amateur attempt at cultured living and were now in the era of the Berni Inn. These provided the next step up in our search for culinary excellence. Their premises were tastefully decorated in dark red flock wallpaper with artificial oak panelling, whilst lighting had graduated from the wine bottle to individual plastic Tiffany type lamps on each table. The whole set-up was designed to create an ambience of both opulence and sophistication, or so we thought. We just had to take Bill.

On arrival we hung our coats on the hooks thoughtfully located behind the door, and sat down. We were presented with one laminated menu between us, listing such tempting delights as Chef's home made soup, Duckling *a l'Orange*, or, (for the very daring) there was Gammon Steak with a slice of pineapple on top, followed by Black Forest Gateau. What more could our American guest wish for?

Interestingly, the soup 'of the day' was invariably reconstituted Minestrone (to remind us of Italy) but would have been hand stirred by the chef himself.

The main course was always listed as being 'Accompanied by vegetables in season,' but wherever we went the only 'in season' vegetable turned out to be frozen peas.

However the greatest advance was in the choice of wine. We now received a 'Wine List' of some five or six wines and were able to exercise our newfound knowledge by casually choosing vintages such as 'Blue Nun.' What a long way we had come in such a short time.

One thing we hadn't bargained on however was Bill's compulsion to 'chat-up' the waitresses. We watched in mounting horror and embarrassment as he began his patter, expecting at any minute to be thrown out for sexual harassment. Not a chance; the object of Bill's attentions was lost to such an extent that all thoughts of waitressing were forgotten.

We now saw at first hand the awful truth—even without having to brandish nylons, our transatlantic cousins were still able to reduce the average waitress to a blushing teenager. And it was happening here in a Berni Inn in England and it was my fault.

Sadly, there was one occasion when Bill's compulsion to charm failed him and he cracked. This must have been our fourth meal out, and by now we were all used to the sequence of events.

As usual the waitress approached and Bill went into action; 'Hello honey, my you look fabulous. That's a gorgeous uniform; it sure shows off your figure. I'll bet you have lots of admirers,' - on and sodding well on.

Soon the waitress was in an advanced state of meltdown, all thoughts of asking what we wanted to eat having long since vanished. She was now warmly wrapped in a gorgeous Californian accent, talking to her just the way she had seen Robert Redford talking to Barbara Streisand.

At this nauseating point, I spoiled it all by asking about food.

Pulling herself together and glaring at me, she reeled off the menu, which again included 'Vegetables in season.' Bill, still in film star mode, asked what the vegetable was.

'Peas.'

'What, peas again?' he cried. 'Don't you Brits eat nothing but Goddamn peas? Ever since I arrived we've had big peas—small peas, fat peas, cold peas, hot peas and some appalling mess called mushy peas. Have you never heard of carrots for God's sake? I am sick and bloody well tired of blasted peas, so don't give me any more of that crap.'

We all stared at Bill, the waitress nearly fainted and he carried on muttering about the massive vegetable selection he could have back in the land of the free.

It just goes to prove how different we British are to the Americans when it comes to stoicism.

Obviously, during the last thirty years, we have become far more sophisticated and nowadays the vegetable of choice in restaurants is, of course, *'petit pois'* and, if we are talking in real gourmet terms, they will be served with a sprig of mint on top.

If only Bill could have stuck around for longer.

10

And so to Work

Now, as everyone knows, the work ethic is a cultural norm that advocates being personally accountable and responsible for the work that one does and is based on the belief that work has an intrinsic value.

Well, as you have seen so far, we in the offshore scene certainly subscribed to that ideal. At least, I think we did. Having dealt with crucial issues like films and newspapers, we now need to turn to rather more mundane matters such as keeping the platform operating. (If this were a technical document I would have added the word 'safely' to the last word. However, it isn't, so I haven't).

I recall visiting a platform under construction in a Fjord (a big, deep, cold pond in Norway) and being somewhat taken aback to discover that the walkways were ankle deep in assorted electrical and gas cables. Even more interesting was the fact that they were all lying in various depths of water, with sparks flashing and bubbles blowing, as I walked, very tentatively, over the spaghetti.

On querying the safety aspects of the situation with a Supervisor, he answered, somewhat tersely, that as they were behind schedule, they weren't ready to fully implement safety yet. Being sensitive to nuances in language I realised this was a coded message for me to 'Sod off and mind my own bloody business.'

Now, an average shift on board started at 7am and finished at 7pm. During this time, as we weren't yet producing oil, we had Drillers drilling wells, Constructors hitting things with hammers and the production group getting to grips with the intricacies of the process plant and equipment.

My maintenance gang already had work to do on the life support systems, electricity was being generated, fresh water was being made, carpet tiles were being 'hoovered' and the galley was making chips and custard (not necessarily at the same time.)

As far as *we* were concerned, the most important work was underway.

One mammoth task we had early on was to 'Trace the pipework.' The oil process and utilities plant was connected by literally miles of assorted pipework, in a dirty black or rusty brown colour. It went from one end of the platform to the other, through walls, floors, and ceilings and up and down the sides of the jacket. As this had just been constructed, none of the pipe was colour coded, so we had no idea what was supposed to be carried where and by what.

'Tracing' involved teams of technicians from maintenance and operations, who crawled all over the platform, checking drawings and following each pipe from start to finish. As we progressed, colour coded identification stickers were secured to each pipe with an arrow to show the direction of flow.

Now, at any given time, it was not unusual for the maintenance team to be feeling their way along until, like the builders of the Panama Canal, they met the production team coming the other way, as each had begun from different ends of the same pipe. This in itself was no problem, unless one team was convinced that the pipe in question was an oil export line whilst the other was equally insistent that it was for firewater.

Things could become somewhat fraught when one team finally realised they must be wrong and were faced with backtracking, with the added fun of having to remove all the identification bands fitted so far. It was an interesting reflection on human nature that initially the discussion concentrated on the need to 'win' the ensuing argument, whilst the actual contents of the pipe were largely ignored.

The exercise helped to develop a healthy understanding of respective positions, articulated in the view of Operators that 'Maintenance are a waste of space,' and in the opinion of Maintenance that 'Operators are a bunch of no hopers.'

Fortunately, I had become good friends with the American Production Supervisor. Since we were blessed with intellectual superiority, we were able to stand aloof from our respective teams' tantrums. This healthy relationship was due in no small way to his laid-back attitude to conflict and a baffled acceptance of my having installed scaffold planks in our bedroom.

I suppose it would be called 'management style' today, but at that stage we didn't know about such things, as we hadn't yet been exposed to the dubious benefits of Personal Development Courses and the ministrations of Total Quality Management Gurus..........

Talking about management style, when I arrived on the platform, construction was still underway. One of the biggest jobs in progress was the lifting and fitting of the flare boom at the South end of the platform.

Lifting the boom and fixing it in place was done by an International Contractor working from a giant Crane Barge moored alongside. They employed mainly foreign labour and at one crucial point the flare boom was suspended some 280 feet above the water while a team of Spanish welders hung on to the swaying structure, in an effort to secure the boom to the platform.

One of our Engineers told me that in the week before I arrived, one of the Welders had lost his grip and fallen off the boom. Shocked at the news, I asked whether he was all right.

'Oh, he was fished out okay, but his Supervisor was quick off the mark and managed to sack him before he hit the water.'

The flare stuck out at an angle of about thirty degrees and was about eighty feet long. It was built of tubular steel with the gas pipe running up the centre. There was a steep narrow walkway alongside the pipe and when you climbed up to the tip of the flare, it moved up and down in time to your footsteps. This was great for your confidence, as the rocking motion did its best to throw you into the sea; however matters got much worse when another idiot decided to climb up behind you to admire the view. He, of course, would be blissfully ignorant of the growing oscillations his footsteps were inducing at the tip and the increasing hatred and panic of the guy at the end, who was soon hanging on for dear life. Upon such rocks are old friendships dashed.

As time moved on we found that most of the heavy maintenance work was of a mechanical nature and so we formed a small team with me nominally in charge, and Liam and Bert taking no notice.

As I said earlier, Liam was already familiar with offshore life and so Bert and I were, for a time, foolish enough to rely on his superior knowledge. Liam was a very interesting chap and had many talents, not least of which was he was a natural crane driver.

This may sound mundane, until you consider that we had two cranes, one on each side, which were mounted on forty-foot high steel pedestals. Each crane had a boom length of 110 feet and was capable of lifting some twenty tonnes. They were powered by huge diesel engines which also acted as counterweights.

The cranes were in continual use lifting containers and materials off the back of supply boats and moving heavy drilling equipment across the deck. As the Drilling Department lived in the mistaken belief the cranes belonged to them, they also provided the official crane drivers.

Liam's skill in crane handling lay in his unerring ability to pick up containers off a supply boat even when there was a twenty-foot swell. This meant that the timing of the lift had to be judged precisely. If you lifted too soon, the boat would rise on the crest of a wave and smash into the base of the container, causing the crane rope to go slack. At this point, the load would either crash back onto the deck or, wave about like a giant pendulum. Obviously, such mistiming would endanger the lives of the boat crew who, being on an open deck, had nowhere to hide.

Liam's fame spread to such an extent that in boisterous weather it was not unusual for supply-boat Skipper to ask that Liam take over from the Drilling crane driver. This really endeared Liam to the regular drivers, something he affected not to be aware of. We, on the other hand were rather proud of our 'lad.'

He really enjoyed looking after the cranes. One day when discussing among the three of us what, in the mysterious world of maintenance, we should do next, Liam piped up, 'The West crane rope needs to be changed.'

All unsuspecting and being managerial, I replied 'Should I bring out a crane specialist?'

Liam gave me one of those looks that makes you realise you have just lost what little credibility you had, and said,

'It's a doddle; the three of us can do this, no problem.'

With words like 'doddle' being bandied about, who was I to argue?

So, later that day, as instructed, I found myself inching (and I do mean inching) along the top of the crane boom to the very end where the rope went over a huge pulley and down to the hook. From here I also had the dubious pleasure of looking down onto the surface of the water some 140 feet below

What Liam had been careful not to mention was that being 110 feet long, the boom responded to my progress by moving increasingly up and down, (reminiscent of the flare boom, but without the added comfort of a handrail.) Seen from the deck this was hardly noticeable, but as experienced by a nervous idiot crawling towards the end, it was leaping about something awful.

We then had to unwind the old rope, whilst at the same time winding on the new rope, which was coiled on a big drum on the deck.

My role was to guide the rope over the pulley, Bert's role was to feed the old rope off onto the deck and of course, Liam was in his favourite place sitting inside the crane cabin, all warm and cosy, handling the controls like an artist.

Meanwhile the boom continued to rock up and down, I continued to hang on and the sea continued to thrash about a very long way below.

What Liam had also neglected to mention was that the new rope, all three hundred feet of it, was liberally covered in a hell of a lot of grease and

48

slowly but steadily, most of it was being transferred onto me.

Part way through the exercise, I distinctly remember experiencing the classic 'mixed feeling' syndrome—If I survive I will definitely sack Liam when we finish, and—Why on earth had I left a job in a nice warm classroom?

These unworthy thoughts faded away when, slipping and sliding off the crane, we reluctantly agreed that Liam's scheme had worked. Our reward for this endeavour was the usual strong brew and sticky cake, in our office.

One small but noticeable niggle was that Bert and I were covered in rapidly congealing grease, while Liam was still nice and shiny.

It did, however, teach me a lesson regarding the importance of interrogating Liam carefully when he 'had a good idea,' just to preserve the illusion that I really was in charge.

Before we leave the cranes, there was another occasion when Liam approached me with a sound but lunatic proposal. He wanted to remove the entire engine from the East crane and send it ashore to be repaired. Apparently, the engine's performance was suffering and Liam, who nursed them like babies, was convinced that if we didn't act immediately, the engine would be a 'write off.' This, he added, would require the Company to purchase, at great expense, a replacement unit from the USA, thus causing considerable delays to our drilling programme. There's nothing like a bit of understatement to convince waverers.

We examined the engine and I agreed we should proceed. Liam's eyes lit up, bags full of spanners appeared and he and Bert set to and removed the engine by lifting it out with the other crane. (I told you he was an artist.)

By about 8pm that evening the East crane's engine was swinging gently from the hook of the West crane, waiting for a supply boat to arrive and take it away.

In the meantime, I had sent a telex to alert our Office that one knackered crane engine was on its way for repair. I was having a nice cup of tea in Sid's office, when all hell broke loose.

We became aware of a tremendous commotion in the corridor and on looking out I saw the Toolpusher rushing towards me, brandishing a spanner, with murder in his eyes. Not having a clue what he was on about, I retreated behind Sid's desk just as he burst through the doorway.

This was alarming enough, but what made it worse was that his ranting was totally unintelligible, as he was speaking Dutch, with an Indonesian accent. By now quite a crowd had gathered in the doorway to see the fun, as he continued towards me. The incomprehensible diatribe was now accompanied by banging his spanner on the doorframe, a filing cabinet, a swivel chair and finally on Sid's desk.

Sid, ignoring for a moment the imminent destruction of his desk, took control. 'You sod off out of it,' he said to me, 'While I find out what's upset this lunatic.'

I obeyed immediately, and dodging the madman and his spanner, left the room (as they say.) Then, with great presence of mind, I sat ever so calmly in my office with the door firmly closed.

Later, there was a knock on the door and Sid's voice intoned, 'It's okay, you can come out now, I've calmed him down.'

As if I had been hiding.

Apparently, the Toolpusher was upset because he wanted to lift some drilling equipment with the East crane, but unfortunately he couldn't, as the engine was now hanging in the air and swaying gently in the breeze.

On discovering the awful truth, he apparently went berserk and demanded to know who was responsible for ruining his day.

And some kind soul told him it was I.

Never having been exposed to a real psychopath before, I found the Toolpusher's behaviour somewhat baffling, until Sid explained that he had spent most of his career on drilling rigs in the tropics, very far from civilisation, and had obviously been badly affected by the humidity.

In those remote locations, he was looked on as some kind of god, whose word was absolute law. What was disconcerting, as far as we were concerned, was that he still believed he had every right to dispense summary justice on the natives, in this case–me.

Sid said the man just couldn't come to terms with the fact that the platform was not his own giant Rig and that he now had to share it with a bunch of ill disciplined layabouts like us.

So, two more lessons learned; working with Liam could get you killed by mad Toolpushers and to beware of jobs with high grease content.

Just a couple more things they don't teach you on McKinsey Management Courses.

Liam was great tinkerer and something of a 'radio ham' when at home. Knowing this, it seemed perfectly reasonable to find him one evening, clutching a handful of cheap wire coat hangers. When I asked what they were for he replied, 'I reckon I can make a TV Aerial with these and then we'll be able to get decent reception.'

Now, at the time, although we had a television, we couldn't get a signal from onshore. Most of us didn't mind in the least not having TV for a week or two and to be honest, neither did Liam; what drove him on was the irresistible need to tinker.

We had already shipped out a TV specialist who, having failed to get a signal of any sort, had pronounced it to be impossible, due to the curvature of the earth. As this was by far the most impressive excuse we had heard so far for failing to achieve anything, we accepted his explanation and reconciled ourselves to missing essential viewing like *Ask the Family*.

No sacrifice too great for a good Company was our motto.

Anyway, back to Liam, to whom this explanation was totally

unacceptable. That evening casual watchers observed a discreet hole being drilled through the accommodation wall, a length of co-axial cable being fed through and carried up the outside for two storeys to be tied off on the outside rim of the helideck.

Liam now recruited Bert and I as acting unpaid technicians. Would we never learn?

Bert and several other secondees were positioned at intervals down the stairs while I was delegated to watch the TV screen for signs of life.

The coat hangers were fashioned into a design unique to Liam with pliers, accompanied by the odd curse as sharp ends ripped his overalls and caused lacerations. At last he was ready to start and began wiggling the aerial about in various directions, shouting 'Anything yet?' to me via the chain of assistants, only to have the message 'Not a sodding thing,' beamed back.

This went on until the 'assistants' became disillusioned, compounded by the realisation that they were missing a crucial game of snooker in the warmth of the accommodation. Slowly the volunteers started to drift away from the advanced electronics undertaking, despite Liam's accusations of disloyalty and lack of stamina. The inevitable result was he found himself alone in a growing gale, a condition which would be repeated on a number of occasions.

After this, Liam said he wasn't about to put himself out any more for such an ungrateful bunch and thankfully no further attempts were made to emulate John Logie Baird.

Probably due to the disappointing outcome, the aerial was left fastened to the underside of the helideck. One day, we were having a Certification Body inspection and the Inspector, noticing the assorted coat hangers waving in the wind, asked what they were doing there. Thinking quickly, the Helideck Supervisor replied that he believed it was something to do with the radio room and the need to boost a signal. It's on occasions such as this that one is proud to work with people who can lie with such consummate ease.

One day, at his morning meeting, and by way of showing he had his finger on the pulse, Sid the OIM said, 'I see we're welding at the far end.'

We all looked at each other in some puzzlement, until the Construction Supervisor said, 'It's okay, it's us, and we've found the perfect place for our fabrication and welding shop.'

This statement seemed to baffle Liam, so he muttered, 'I'll get the drawings out.'

Bert said, 'There's no need, I know where they are. They're in the storage area next to the flare, but I can't figure out where they found room, as the floor is full of our chemicals.'

'Yes that's where we are, but we've found a grand place on the mezzanine level, so the lads have tack welded some old stairs to the side. There's a lovely big space on top. It's just ideal for us and we're making a bench so we can weld stuff without kneeling on the floor.'

And with this example of initiative at its best, he sat back, waiting for our congratulations.

Sid, who hadn't said a word for a while, nodded as though it all made admirable sense and commented, 'I could see the flashes through the vents in the upper wall while I was having a stroll, what is it you're welding?'

At this point our attention was diverted by the sight (and sound) of Liam in some kind of inner turmoil.

'Mezzanine, that's not a bloody mezzanine! We don't have a sodding mezzanine, what you morons are welding on is the top of the main diesel tank.'

'What?' I think we all said, as one voice.

'Liam's right,' said Bill, 'It's got to be, there's nothing else in there.'

'Is there any diesel in the tank?' said Sid, coming over all managerial.

'It's full,' I replied. 'Don't forget that until we get gas flowing, our three gas turbines are set to run on dual fuel and there's also our stand-by generator which runs solely on diesel.'

'Bloody hell!' exclaimed Sid, which seemed to sum it all up.

Attention now turned to our friend the construction supervisor who had been conspicuous by his silence during this exchange of information.

'Well I didn't know it was a tank,' he said, 'And anyway we've hung a fire extinguisher on that pipe running up through the roof.'

I think it's called trying to retrieve the irretrievable.

Bert said 'That pipe you're talking about is the diesel vent pipe. It will be nicely full of fumes, so your extinguisher could well come in handy, although it would be about as useful as throwing a handful of sand into a volcano and shouting "Stop it Vesuvius."'

Bert had a nice turn of sarcasm when he put his mind to it.

Sid, having stopped shuddering, said, 'Sod off and get that lot off the tank and don't use your initiative ever again while I'm on board.'

I thought, 'Nice one Sid,' even if the order seemed to be a little selfish in scope.

After our pyrotechnics maniac had fled the scene, we sagged back in relief at a disaster averted and Liam said, 'I think being subject to all those flashes fries a welder's brains.'

'Either that,' said Bert, 'Or they're wearing spotty hats that are several sizes too small.'

'Oh, and Sid,' said Liam, 'Those vents you saw the flashes through are part of the blast wall, so had something gone wrong, there would have been an even better view, as it disappeared into the sea.'

'Don't tell me any more,' growled Sid, 'I've had enough bloody excitement for one day.'

There seemed little else to say after this in-depth analysis of the situation, so we all went for a cup of tea.

In the dark ages, before satellites were invented, we still had to talk to other installations, helicopters, boats and, reluctantly at times, people onshore.

The focus of all communications activity was the Radio Room, which was located in the accommodation. Casual access was discouraged and tight security maintained by the use of a home made stable door, with the top half open and the lower portion having a shelf for leaning on.

Ken our Radio Operator was in absolute control of the inner sanctum, primarily because nobody had the remotest idea what all the humming, bleeping, flashing equipment was for; and because he took great care never to explain how anything worked.

In the eyes of most people, Ken wielded awesome power, being the only man on board who knew when the crew-change helicopter was due. This fact alone ensured he had a good seat at film shows and a place always available at the dinner table.

There were two main ways to make contact with the outside world; the first was to use 'Telex' which was a means of handling telegraph messages in the form of coded radio link signals. Telex was widely used, as it was constantly available and as a result, was the means by which written reports, instructions, requests and crew change lists were transmitted back and forth.

Again, it is worth remembering that there were no personal computers on board and all reports, etc, had to be handwritten on a telex pad then typed by Ken onto the teleprinter, prior to sending to the recipient. This sounds easy, until you realise that the Drilling, Construction and Production daily reports had to be sent overnight, so it was not unusual for Ken to work a strange twilight shift of his own devising.

The one he hated most was the Drilling telex, which often filled a seemingly endless sheet of paper some twelve feet in length. It contained lots of technical gibberish and random numbers and on receipt was cut into 3-foot lengths, mounted on a wall and stared at by the Drilling Department onshore.

This enabled some brainy, but sad, individuals to make decisions regarding the contents and send an equally incomprehensible reply back to the platform. If we were to judge competency by yardage of paper, Drilling would have won every time.

The second method of communication was affectionately known as 'Wick Radio.' This was originally an Admiralty Station, opened around 1907 and located at the northern extremity of the UK mainland. Over its lifetime, it was involved in many distress incidents involving ships in some of the world's most inhospitable waters. The station was also the main link for the UK's deep-sea trawler fleet and provided a Morse telegraphy service, keeping them in touch at the fishing grounds off Norway, Spitzbergen, Iceland, Greenland and Newfoundland.

Speaking of Morse, many radio operators were recruited from the Merchant Navy and were experts in sending and receiving Morse code. I remember one evening watching Ken sending a Morse message to a colleague on a passing ship and asked him why he still used the system, now that we had voice contact.

He replied that he liked to keep in contact this way and that Morse was not subject to signal distortion in the same way that radio signals could be. You learn something every day.

Using Wick Radio required considerable planning and discipline, as there was only one telephone, which was located just inside the Radio Room. Contact was managed, once again, by Ken (I told you he held absolute power) and the system worked roughly as follows:

To make a call to shore, you entered your name on a list posted outside the radio room, bearing in mind that all the Department Heads invariably wanted to call their Bosses at the same time. Whilst most of us learned to accept this minor irritation, it was a lesson in democracy that caused apoplexy on the part of the Toolpusher, who still couldn't come to grips with the fact that he was no longer the supreme ruler of the airwaves.

Once your name was on the list it was a case of waiting for Ken to call you when it was your turn.

Now, a minor problem was that Wick Radio was not really geared up for this vastly increased demand for its services, so in order to cope, they set up a strict regime whereby your available contact time was restricted to 10 minutes. This rule was observed to the letter and it was not unusual for the voice of the Operator to interrupt your conversation with 'Hello Platform, your time's up,' and you were simply disconnected. You soon learned not to waste time asking about the weather, football results, the latest on Coronation Street or whether Harold Wilson was still the Prime Minister.

There was also one other critical discipline to learn, my only experience of which was confined to war films. It was the need to recognise that only one person could speak at a time - which sounds easy until you try it.

You had to grasp the fact that you couldn't reply until your opposite number said the magic word 'Over.' This then enabled you to speak, but obviously you had to remember to say 'over' in turn if you wanted a reply. Many's the time, that in order to say something of earth shattering importance, a person would rabbit on for a while, only for Ken to say 'The other side didn't hear a word of that—they forgot to say "over."'

This was made even more enjoyable when, having finally got an 'over' the sender tried to repeat the urgent message, only to be cut off in mid sentence by Wick Radio. What fun we had.

Even more exasperating was to find, on finally getting your turn, that the person you had been waiting all morning to confer with had gone to lunch. Funnily enough, in later years, when communication was all too readily available, I sometimes hankered for the good old days when you could depend on Wick Radio to break contact at what, in hindsight, was an opportune moment.

We had a one period of utter confusion when we found that one of our American friends, who was already used to the one way system, said 'Come back,' instead of 'Over.'

This doesn't sound too problematic until you realise that he hailed from the deep south of the USA and his rendition of 'Come back' bore no phonetic resemblance to any words that we recognised. His southern drawl became even more interesting to listen to when he was talking to the Skipper of a supply boat out of Lowestoft. The dialogue went reasonably well until the 'Come back' phrase.

In an equally incomprehensible Suffolk dialect, the Skipper kept asking our American why the hell he wanted him to come back just as he was about to sail. Our colonial friend then tried patiently to repeat the call sign in something resembling English. As mutual frustration grew, it became obvious that neither one understood a word the other was saying, and both resorted to shouting in an attempt to communicate. It's that sort of thing that really brightens your day.

Speaking of bright days, there was a particular event to which I as a member of Ken's inner circle was invited into the radio room to witness. As I say, all radio communications were relayed through Wick, but what I may have omitted to mention was a small but interesting point; if you tuned your receiver to the same frequency, you could listen in to other people's conversations. MI5 should be so lucky. In general, this was boring in the extreme, but there was one significant exception. Ken had discovered that an American Barge Engineer, working just inside the Danish Sector, telephoned his wife at the same time each Sunday morning and quite accidentally, he had started to listen in.

Being a kindly soul, he felt it only right to let us share these conversations, purely in the interests of expanding our telecommunication knowledge, you understand.

So, there we were one Sunday morning, Sid, Liam, Bert and me, crowded into Ken's lair, waiting for the drama to unfold. Now, you may be shocked at our voyeuristic attitude, but remember, we had no newspapers and the films were terrible, so what options did we have?

Ken tuned in to the correct frequency and the conversation began. To say that it was 'one sided' would be an understatement. You see, the Barge Engineer knew there was a distinct likelihood that others may be listening to his conversation, but sadly for him, he hadn't got this crucial message through to his wife. Initially, the conversation was fairly mundane; things like 'Hi honey, how are you?' and 'Did you get the bills paid?' were quickly dealt with.

Then, to our eagerly anticipated delight, his wife would start to wax lyrical about their love life, how she missed him and what she would do to him, on him and with him when he returned. Then, having exhausted herself (and us), she started to ask him how much he loved her and how would he prove it on his return. At this point, the husband would become completely monosyllabic in his responses, suspecting, quite rightly, that he had an attentive audience.

All he kept repeating was a long drawn out 'Yup,' to every amorous query from his wife. However, as regular listeners to this saga of everyday love-life knew, these responses, as proof of reciprocal ardour, were, as far as she was concerned, somewhat unsatisfactory. Failure on his part to become the Ann Summers of the airwaves, led to her pressing for even more reassurance that 'Booboo still loved Kitten,' until finally the voice of Wick Radio called time. By now we were all helpless with laughter and convinced we could feel the heat of our friend's embarrassment across the airwaves.

We were just about to go for a restorative mug of tea when Liam startled us by saying that the ungrateful sod didn't deserve a gorgeous wife like that; and furthermore, had he been married to her, nothing would make him work offshore.

We felt this to be somewhat out of character for Liam, until Ken reminded us that he had only ever had one call from his wife. Obviously we needed to know the details and Ken said, 'As I remember, her *amour* centred around a flooded kitchen and what was he going to do about it?'

Apparently at no time had Ken heard Liam's wife ask whether 'Booboo loved his Kitten.' For some obscure reason, Ken's revelation seemed to upset Liam even more and we had to calm him down with a sticky bun.

Sadly Wick Radio closed in June 2000 and there are now no Coastal Stations in service. Staff employed were probably at their highest levels during the two wars and in more recent times during the 1960's and 1970's dealing with increased traffic from fishing boats, supply boats, diving vessels, rigs and platforms.

The UK's Morse Telegraphy service had been finally closed in the early hours of the 1st of January 1998. The time chosen for closure dictated that there could be no on-air 'jamboree' of farewells. However, Land's End Radio managed to send a final broadcast:

'But now the time has come,
ours is not to reason why,
the satellites are calling,
our Morse transmissions die.'

The broadcast ended with:
'Marconi, if you can hear us, we salute you'

Several faxes were later received from ex-seagoing Radio Operators who had taken the trouble to listen to the closure broadcasts and expressed their thanks for the work of the service over the years.

And so ended a proud era, due as they say, to the march of progress.

I read somewhere that 'Communication involves both the giving out of messages from one person and the receiving and understanding of those messages by another. If a message has been given out by one person but not received or understood by another, then communication has not taken place.' Who sits down and comes up with these gems and why?

Now I realise that idly quoting the above is not something guaranteed to get you free drinks in a pub; in fact it may even get you thrown out of one. But I was soon to find that the ability to communicate was to test me in a way I had not expected, as we will see in the next episode.

11

Ship Ahoy!

'Maintenance Supervisor to the radio room.'
Liam and Bert were alternately banging, hammering and swearing at the Water Maker Unit, as you do when a delicate, highly technical maintenance operation is underway. I was at their side pretending to be in charge, when the tannoy repeated, 'Maintenance Supervisor to the radio room.' As this meant quite a climb back to the accommodation, I tried to ignore the message, until Liam said it might be something important, like a delivery of new films.

I had obviously failed to recognise the urgency of the message; why else would I be called? So off I went. When I arrived at the radio room, Ken said 'There's an urgent telex from your lord and master onshore, which I thought you might want to see.'

Somewhat puzzled, I took the telex and read, 'Tanker Trials to commence on Thursday, Maintenance Supervisor to take charge of the exercise. You will liase with the ship's Captain and supervise mooring to the Buoy. Message ends.'

To say I was taken aback would probably be something of an understatement.

I told Ken there must be some mistake in the addressee section, and he did my confidence no good at all by saying he thought so too. Believing it was a mistake, he had contacted the telex room onshore, which confirmed the accuracy of the message. My only hope now lay with Sid; surely, he would be able to rectify this obvious error.

I waved the telex at him, protesting that only two months ago I had worked in the mining industry where super-tankers didn't figure much in our daily routine and that my experience of ships to date was limited to trips across the Mersey on the Royal Iris.

Sid said he knew about the message, as he was copied on the distribution list and without further word, he delved into his desk drawer and handed me a package, saying, 'I guess you'll need these.'

Cautiously opening the 'gift' I was horrified to find a massive pair of binoculars, as used by Jack Hawkins in the Cruel Sea. As I stared at them, Sid said, 'You've nothing to worry about, the tanker Captain will know what to do.' Another boost to my already non-existent morale. Oh and by the way, this was Tuesday, so there were two whole days to go before I did my Admiral of the Fleet act.

Now, before going any further I should explain how oil was transported from the Platform to the Refinery. Although we weren't yet in production, all the big construction work had finished and we were concentrating on fine-tuning the process facilities. Drilling had completed a number of wells and checks were being carried out on emergency shutdown systems. So tanker trials were next on the pre-production agenda, and as there were no storage facilities on the platform, we could only produce oil when a tanker was available. In order to make this possible we had installed a special loading Buoy about a mile away from the platform, which was connected to us by a pipeline on the seabed.

The Buoy resembled a fishing float of gigantic proportions, with a draught as deep as about three cricket pitches. The submerged portion contained a central shaft with the surrounding area filled with water ballast and buoyancy compartments. Massive four-tonne counterweights hung down the shaft and were used to counteract the pull of the tanker on the oil hose and mooring rope. The entire unit was held in place by eight anchors on chains, which extended outwards from the structure for about two hundred yards and served to minimise roll and heave in bad sea conditions.

The superstructure of the Buoy could rotate around 360 degrees, which enabled the moored tanker to 'weathervane' through a complete circle and adopt a position of least resistance to the combined forces of wind, waves and current. A helicopter could land on its ninety-foot deck, (a not insignificant facility as we will see.)

Under this helideck were the fifteen-foot diameter reels for coiling the hose and mooring line, a long mooring extension arm, an emergency generator and very basic accommodation capable of housing up to three unfortunates in great discomfort.

The structure above and just below the waterline was protected by a circle of huge wooden fenders, designed to collapse in the event of a tanker drifting into the Buoy. All very reassuring.

Sorry about the technicalities, but the main thing to remember is that it was remote; it had very little protection from the elements and was generally very cold, even in the summer. Oh, I nearly forgot, it was painted in a tasteful bright yellow, so from a distance it looked like a banana with a big nose wearing a wooden skirt and a flat red hat, floating in the water.

We now come to the tanker mooring part of the saga.

Obviously I wasn't about to undertake the tanker trials on my own, so calling into play my managerial skills, I 'volunteered' Liam and Bert to form a close-knit, largely incompetent tanker mooring team. Having established that the tanker had set sail and realising that our prayers for a typhoon were unanswered, we were flown out to the Buoy in the Bolkow, which having deposited us, disappeared back to the platform.

So there we were adrift in the North Sea wearing insulated overalls, helmet liners and gloves. We each had a Motorola radio, a robust unit carried in a leather case, which although cumbersome by today's standards, was a reliable piece of kit and made us feel important.

As befitting my superior status, I wore both the marine binoculars and the radio crossed over my chest in the manner of a Mexican bandit.

The proposed sequence for mooring the tanker was the sort of thing that must have seemed like a good idea in the warmth of an office onshore. We however, were about to test the system in the middle of the North Sea with a wind blowing at Force 4.

I know Force 4 doesn't sound much, but you try it standing on a heaving helideck with no handrails some forty feet above the water, waiting for an oil tanker to appear over the horizon. We knew vaguely what the plan was, although those who had conceived it had been careful not to provide too much in the way of detail, and refused to answer their phones.

I should now introduce a key player about whom you will hear more as time and mooring operations became routine. We had at our disposal a supply boat that sailed out of Lowestoft to help us moor the tankers. As we were to learn, there would be numerous occasions when we would have been in considerable difficulty without its Skipper's superb seamanship. Over the coming months we became good friends, although we could only ever saw him on the far-off bridge of his vessel.

Our Radio Room informed us that our tanker was about one hour away and we soon should be able to see her on the horizon. Of course, while we waited for her to appear, both Liam and Bert wanted a turn with the binoculars so that they could play at being the Captain of a Destroyer guarding a Murmansk convoy.

Honestly, it was embarrassing to see grown men behaving in such a childish manner. Besides, I had already made it clear that I was the Captain and they were just crew. It looked as if mutiny was about to break out when Bert, who had the glasses shouted, 'I can see her!' and we all calmed down—and then suddenly realising what we had to do, panicked in true Gene Wilder style.

Time to get ready.

In order for the tanker to couple onto the Buoy we had to feed out what came to be known as the 'Messenger Line.' This was a massive polypropylene rope, one end of which was attached to the actual mooring line while the rest was coiled like a huge snake on the deck of our loyal supply boat, which then began to steam slowly away from the Buoy. As she did so her crew paid out the 1000 yards of rope so that it floated in the water. We now waited for the tanker to come into range.

As the vessel was approaching very slowly, it looked reasonably small for quite a long time. However, as it came nearer we began to realise that a 40,000 tonne tanker was a hell of a lot bigger than we were. Bert suggested looking through the wrong end of the binoculars, but we were too nervous to try. As the monster drew closer, we remembered all those physics lessons about the momentum of a ship being so massive that it would be impossible for it to stop on the spot. This made us hope that our physics teacher was talking rubbish—but more importantly, we hoped that our as yet unseen Captain had also read the same book. It's not, you understand, that our faith in his seamanship began to waver—not at all.

By now, even though the tanker was still about a half mile away, it towered above us and both of my erstwhile assistants began to mutiny. Fortunately at that moment, the tanker called us on the radio, wished us good afternoon and asked whether we were ready to begin the mooring operation. Oozing a confidence we didn't feel, we said we were. The next phase was a tricky manoeuvre, made even more interesting because it had never been tried before.

The tanker had to steam slowly alongside the messenger line, catch it with a grapple on a long rope and then, tying the end onto a specially constructed winch, start heaving the line in. The difficulty they initially experienced was in trying to grapple the messenger line from the bow, which, as the ships tanks were empty, was very high out of the water.

Their first attempt to catch the line failed and so the tanker had to move off, go around in a huge circle and come slowly in again. The wash streamed along her sides as the Captain lined her up and once again came towards us at about three knots. This time, they managed to pick up the messenger line and we let out an involuntary cheer—most embarrassing.

The ship's winch started to reel in the line while the Chief Officer, who was standing in the bow overseeing the operation, maintained radio contact with the Captain, calling out distances as they came ever closer. Slowly, at our end, the mooring rope unwound off our reel whilst pulling the counter-weight up the central shaft. Finally, the tanker was connected to the Buoy and having made fast, her crew reeled in the oil hose in a similar manner and coupled it up to their filling line.

Once coupled up she was just about thirty yards from the Buoy and looked big enough to pull us clean out of the water without even noticing. As both our mooring rope and oil hose were both hanging on counter-weights, the tanker had to keep her engines running astern at a steady thirty revolutions per minute to avoid drifting into us.

After some discussion with the Captain and First Officer, we agreed to carry out the unmooring operation. Fortunately this also went well; the oil line was disconnected and reeled in by the counterweight falling steadily down the shaft. Then followed by the mooring rope being released, again reeled in by its counterweight. Finally the messenger line was fed down from the tanker into the water for the supply boat to retrieve and coil up on her deck, ready for next time.

First trial completed, lessons learned and the bones of a procedure captured. As I said to Liam and Bert, I always thought it would be no problem, a statement which they didn't believe for one second and one which I would live to regret.

It was time for us to call up the Bolkow and return to the warmth of the mother ship, as they say in space odysseys. At the end of the day, we had to confess to a rather smug feeling of satisfaction and there was no more talk of mutiny, although my crew still felt the binocular sharing arrangements remained unresolved.

Oh, and by the way, did I mention there were two identical tankers, which would perform a shuttle service? Mind you, when mooring the next one we could at least pretend to be experts.

Finally the realisation dawned that we had to go into production. Onshore would no longer allow us to just play with the plant and equipment; apparently they wanted us to send them lots of oil.

Directives were issued and Sid received an official telex, the gist of which was that production was to start on the stated day, no excuses would be tolerated and a senior person would be coming out to ensure there was no slacking in the ranks.

On his arrival, plans were drawn up, programmes were devised, work was allocated and schedules were to be met. The whole thing reeked of efficiency—something we were not really used to.

The production vessels and pumps were set out in a long line from one end of the platform to the other and were known as a 'train.' In fact there were two trains, located side by side. This was so that we would operate one train at a time, the choice being governed by the location of the drilling rig.

When the rig was pulled over to the right, we would run the train on the left, the idea being that the weight of one would counter-balance the weight of the other and prevent the platform toppling sideways into the sea.

As the rig was expected to stay on the left side for some time, our task was to get the right hand train ready for production.

We worked round the clock to prepare for the oil to start flowing; a milestone which was reached on 21st of December 1975, making what was romantically called a 'perfect Christmas present' especially for the

Company.

Bathed in the warm glow of achievement and the even warmer glow of the flare, we congratulated ourselves and waited for a visit from our Managing Director. He duly arrived by chartered helicopter and was greeted on the helideck by Sid, who had changed out of his slippers for the occasion.

Sid had briefed us beforehand on the protocol to be followed and a number of us were chosen to meet the great man in the recreation room. Sandwiches and 'wee cakies' were prepared for the event. We were scrubbed clean and Sid made the introductions.

The great man said he would like to say a few words, so we put down our toasting orange juice and an expectant hush descended on the congregation. Referring only to his copious notes, he cleared his throat and congratulated us on a job well done and the fact that we were at last contributing to the Company's profits and his pension. Then came what might be termed as the best backhanded compliment we had ever heard.

He said 'It should be borne in mind that this field was estimated to cost £44 million; however, in the event, it actually cost a massive £58 million to complete. Gentlemen, had we known that beforehand, it is very probable that the platform would never have been built.'

With that he stared at us for a moment and signalled to Sid it was time he was off. We stared at back him, at each other and then at Sid; did he mean the extra cost was our fault? Had we just been bollocked in the nicest possible way? Would we have to make up the shortfall? Did we still have a job?

As I write this, I note that the Scottish Parliament Building was originally estimated to cost £50 million but it ended up costing £431 million. Just what our great man would have made of this minor accounting blip is difficult to guess. I was talking to Sid later and commented how privileged I felt to see such a fine example of compassionate management at work. Sid simply replied 'That's an oxymoron.'

A short time later we were at maximum production and sending some 40,000 barrels of oil to the tanker each day. This is enough to fill over 10 million green plastic watering cans, each day, every week, all year. Slowly, this crept up to 45,000 barrels and lots of hand-rubbing was going on by Accountants. Otherwise known as 'bean counters' these are people who create solutions to monetary problems you didn't know you had, in a way you don't understand.

Sid called a meeting one morning to announce that he had just received yet another very important telex from the powers onshore. The main thrust of the message was that as we now had several completed wells just bursting with oil, we were instructed to prepare the second train for production forthwith.

The production supervisor stared at me in something approaching horror; I knew just how he felt, but how were we to explain the reality to Sid?

'Er, Sid, we may have a small problem,' seemed to be a sound opening gambit.

'Why, what's the problem, just check the system, open up the valves and away we go.'

This was Sid at his technically simplistic best.

'Sid, we can't just open the sodding valves because in order to get train One going we pillaged a whole load of equipment from train Two, including most of the bloody safety valves, half the instruments and three parts of the alarm system, that's why.'

Silence, other than a shuffling of feet, some tuneless whistling and a gentle clearing of throats, ensued.

Then Sid, ever mindful of a career in crisis asked, 'Will it take long to fix?'

'It depends how long it will take the purchasing group to get the spare parts, ship them out and for us to install them.'

'Why the hell did this have to happen on my shift? You lot will be the death of me,' grumbled Sid, coming over all emotional. It struck us that his gratitude for a job well done so far was rapidly dwindling.

However, he rose to the occasion and composed what was probably his finest telex to base, explaining the dilemma in such a way that ultimate responsibility for this enormous cock-up was due to those onshore having expected unrealistic deadlines on his poor beleaguered team.

I don't think they believed it for one second, but realised they were on a hiding to nothing by arguing. And, as far as we were concerned, Sid's rightful role as our guardian angel was restored.

There was however, one question which still needed to be answered.

'Sid, what about the fact that the platform design specifies that the weight of the train being operated has to be on the opposite side to that of the drilling rig?'

'Ah,' replied Sid, 'Apparently a bunch of highly qualified Design Engineers have been busy doing sums on their slide rules and have discovered there's plenty of additional strength in the structure to allow two trains to operate wherever the rig is located. Isn't that convenient?'

I wondered 'Are you confident they've got their calculations right?'

'Of course I am! It's just a coincidence that I've booked my annual leave when you lot get to try it out.'

After a good deal of hammering and banging we did get two trains running and the production climbed to over 90,000 barrels a day, which meant that, as each of our two Tankers could carry 200,000 barrels, they were on a non-stop cycle of filling, sailing, docking, emptying and return.

If you consider that a barrel of oil produces about 20 gallons of petrol, you begin to appreciate the accountants' vision of heaven. We once worked out that one tanker load could provide enough fuel for an average car to travel 160 million miles at say forty to the gallon. (And, as our beloved Chancellor takes 90% of the price of each gallon, you begin to see how, many years later, we could afford to build the Dome.)

It was at a committee meeting just before Christmas that some fool asked if management would consider paying a bonus at Christmas as a gesture of goodwill. His argument was that by then we would have produced something in the region of nine million barrels, despite many setbacks (you can do the sums and change this to gallons of petrol if you like).

Before Sid could rule this idiocy out of court, an animated discussion took place, which was centred mainly on what would be a suitable figure—one that was capable of providing optimum happiness to the crew, whilst at the same time giving management the opportunity to look truly magnanimous.

After some minutes of total confusion and rising excitement Sid brought the meeting to order.

He had, he said, given the matter of a suitable figure some thought. Again, joy was unconstrained; how much would it be, will it be tax-free and can we have it in dollars? All eyes turned to Sid, not a pretty sight, but money was at stake.

Coming over all oracle, Sid said, 'It is only fitting that we should leave such matters to management. After all they are capable of much greater thinking than we are and to ask now for a certain sum may constrain their own view on the size of the package we no doubt deserve.

'I intend therefore to leave the matter of our just reward to their good nature and sense of fairness.'

As nobody really understood this last statement, there was further discussion regarding the advisability of leaving the matter unresolved. However, Sid reminded us that management would no doubt be aware it was the season of goodwill to all men, even offshore men, and declared the meeting closed.

I stayed behind to try and make sure I had accurately captured Sid's pronouncements and asked 'Do you think we *will* get something?'

Sid looked at me almost pityingly and replied, 'There's not a snowball in hell's chance, and between you and me, I'm not even going to ask.'

Making one last try to understand Sid's machinations I said, 'Why don't you think it will work?'

'Because your work is important but not valuable.'

Once again, as I didn't understand what he meant, I gave up and wrote a totally fictional set of minutes.

Production continued to creep up; inevitably we became intent on creating some kind of record. The only problem was that we were now exceeding the maximum design figures for the process and we kept encountering a snag with a thing called the flow control valve. This instrument automatically reduced the flow of oil if it started to exceed the designed amount. We couldn't adjust it as it was already at its maximum setting, so every time we opened up the wells a bit more the blasted control valve would shut down again.

I know, I've gone all technical again but bear with me, it can't be helped.

Obviously when you are producing 95,000 barrels and know you can get the magic 100,000 barrels, you are not about to let a foolish thing like a safety device stand in your way (if you quote me, I will obviously deny I ever said it.)

We were debating this problem in the control room one day when Martin, our Instrument Technician, said he thought he could fix the valve, but it would have to be done unofficially.

It is at times like these that you realise that the spirit of Isambard Kingdom Brunel is alive and well and we should never have given up India.

Martin said he would need a little time, as he had to make a fairly delicate piece of kit which would be fitted to the valve, thus allowing more fluid to pass through. He then startled us somewhat by asking if anyone had seen any thick pieces of wood lying around.

Somebody said there were wooden pallets lying in the drilling rig sack store, so off he went to see if they might be useful for his plan. In the meantime we wondered what he was up to and how could he alter the setting of the valve using something made of wood. We didn't have to wait long before Martin reappeared clutching what looked like a big wooden wedge with a hole in the large end through which was threaded a length of string.

I said, 'Martin, what's that? It looks like a wedge.'

'Yes that's right, it is. I'm going to wedge it between the top of the valve and the connecting rod, so the valve can't close even when it gets a signal to do so.'

Having imparted this gem of high tech thinking he said he was ready to fit the wedge in place.

'Hang on Martin, what the hell's the length of string for?'

'Oh, that's so I can pull the wedge out if we have a power failure and we have to shut down quickly. It's a sort of safety feature.'

Two important things occurred to me at the same time; one was he seemed to have thought of everything and two, there was no way in the world that management onshore could be allowed to hear about our solution to this quest for production glory.

He duly fitted the wedge and production crept up to a magnificent 105,000 barrels. However, Martin was confined to the Control Room throughout our record-breaking endeavours. Each time we shut down, he hurled himself through the door, down a flight of stairs, up a ladder and yanked out the wedge just in time for the valve to return gently to its closed position. Then, when we started up again, he repeated the exercise in reverse.

I may be wrong but I am not sure we could carry out such entrepreneurial schemes today, especially in the brave new world of risk aversion. Looking back, I am still faintly surprised that no one, either on the other shift, or management onshore, ever enquired how it was that we could produce more oil than was physically possible. You can't beat success for applying the old 'turn a blind eye' syndrome.

Later on in the year, as production settled at a new level, tanker mooring became tremendously important. As one became full, she would disconnect from the Buoy and begin a 15-hour homeward run, just as the second tanker took up position some miles away. Surprisingly, things began to go wrong and we had a number of adventures in the coming months trying to keep the Tankers on the Buoy and the Platform in operation.

As the scars have largely healed, I can describe some of the more idiotic events in which the three of us were often (generally reluctantly) involved.

12

What about Jimmy Young?

The first minor irritation we encountered was the loss of a counterweight rope on the Buoy. This was a steel hawser connected to the mooring rope at one end and to a very heavy weight hanging in the centre column of the Buoy. One day, for no apparent reason, the rope snapped at the top and the broken end was dragged down the shaft by the weight. This meant that until it was fixed, lots of barrels a day of oil revenue would be lost.

Management, on being informed of the setback, didn't take kindly to this loss of production and decreed that maintenance should fix the problem 'forthwith'.

I paraphrase here, as the actual words used were somewhat lacking in grammatical niceties. Having 'got the gist' as they say, there was nothing for it but to gather the stalwarts together and 'assess the situation'.

Again we were deposited by helicopter onto the Buoy. We peered hopefully down the shaft but all we could see in the darkness below was the steel rope looking as though someone had taken a plateful of spaghetti and pushed it into a narrow drainpipe.

We realised two things: one was we were on the wrong shift; and the other was that we would have to pull the whole mess out before we could find what had happened to the very heavy counterweight.

Did I mention that the Buoy had been designed by people who aspired to minimalism and had since gone missing? This wasn't too bad except for the fact that they had omitted to provide any of the facilities we would need for such a job.

This time we needed to discuss the problem with our onshore colleagues, in the vague hope they would give the job to somebody else. However, being 'experts' we reluctantly agreed it would have to be us and in order to retrieve the gubbins lodged deep in the shaft; someone would have to be lowered into the depths, hitch a rope onto the hawser and somehow start to retrieve the mess.

It was obvious that, as this was a potentially hazardous job, specialist help would be needed and as there was also a risk of foul air in the shaft, we decided to engage the services of a Diving Company.

They were used to working in confined spaces (so was I, but I used to get covered in coal dust instead of water) and could utilise their own self-contained breathing apparatus if necessary. A short time later we met the diving team on the Buoy; while in the meantime, the lads had rigged up a temporary winch so that we could raise and lower a metal basket on the end of a steel rope.

The idea was that the diver would stand in the basket while we lowered him into the shaft; I just knew my experience of mine winding would come in handy sometime. And as you can see there was no expense spared in our efforts to help.

The divers came equipped with suits, bottles, hoses, assorted tools and an air of mystique. We came equipped with Liam, Bert, gas detectors, tea and buns. The only thing missing on this job as far as the divers were concerned was water.

It was an interesting experience working with the divers, they were extremely professional in their approach to the job and the Diving Superintendent was very protective of his charges. He and I developed a routine whereby all communication was passed from him to me and vice versa. This meant that I gave no direct instructions to the divers and the Superintendent made no attempt to communicate with my lot, which, considering Bert's tendency to lapse into Doric, was a sensible decision.

Looking back, there is no doubt that had we not thought carefully about the potential risks, the consequences for the descending diver could have been severe. In the event, a patient 'fishing trip' by the dive team achieved success and we were able to retrieve the counterweight and make all the necessary repairs. Simple really.

Many thanks were due to the divers who worked throughout in difficult circumstances and displayed a childlike trust in Liam and Bert's winching capabilities.

One part I did enjoy was the change to radio discipline, instead of the usual 'over,' the Superintendent on receiving a message from the diver, would say to me, 'up the diver,' I would relay this to Liam on the winch, he would confirm and raise the diver. Similarly, the message would be 'down the diver' and away we would go again. Sounds boring, but we were all very aware of the need to maintain a formal approach when relaying instructions. Even though, as Liam said to the divers when we had finished, telling the family that he had been working all day with deep sea divers who never actually entered the water, would be something of a disappointment.

Some time later, I was working with a dive team on a different installation, when we had another interesting experience, this time involving a good deal of water and all of it in the wrong place, but more of that anon.

It was several weeks prior to Christmas. The three of us were once again cavorting on the Buoy and conversation centred on the fact that our shift would be on duty during the Christmas festivities. What goodies lay in store for us? Would we have time off for a huge Christmas dinner? Will they send out a decent film? These and other important issues kept us busy.

I forget what we were doing in detail, except to say it involved us in pulling a new section of oil hose off the back of the supply boat and winching it into position on the hose reel deck. The hose sections were extremely heavy and difficult to handle. However, having done the job before, we had a nice little system in operation and each of us knew what we had to do.

As I recall, Liam was once again operating the winch above us, I was on the fender deck guiding the hose over the edge of the structure and Bert was on the reel platform making sure the hose wasn't snagging on the structure.

Everything was proceeding apace when there was a scream from Bert, followed by incomprehensible shouts, liberally mixed with 'effing and blinding.' Convinced there had been a terrible accident, I called Liam on the radio, he stopped the winch and we both clambered up and down respectively, to see what had happened.

It is at times like this that all thoughts of Christmas disappear and a kind of cold feeling develops, as you fear the worst. Up until then, we three had been accident free and somehow, we believed it would always be that way.

Liam and I reached Bert at about the same time, expecting blood, broken limbs or a collapsed colleague, only to find he was on his feet, missing his safety helmet and speaking even more incoherently than usual.

Bert was a local lad and even at the best of times spoke with a dialect similar to that preserved by Robbie Shepherd in his weekly newspaper column. This time however, Bert excelled himself; he was completely incomprehensible. Despite not understanding a word, Liam managed to calm him down and we established that he was physically unhurt.

Slowly, the awful truth emerged. He had been standing a little too close to the winch rope as it went past when unfortunately; it caught on a guide rail. The problem was that being extremely tight, due to the weight of the hose; it twanged like an elastic band.

71

The upshot was that the rope hit him in the face and knocked his hat off. But what was even more traumatic was that it also dislodged his false teeth, which then promptly disappeared over the side. This at least, explained why Bert's protestations were conducted in a mixture of Doric and gibberish.

Obviously Liam and I began to see the funny side and, being sensitive to Bert's misfortune were compelled to make high calibre jokes at his expense.

'Should we send for the divers again?' 'You look better without them,' and so on.

'Phuff, phuff,' spluttered Bert, which we took to mean, 'Let's finish the job and get back to the platform.'

On our return, he went to see the Andy the Medic just to ensure there were no injuries to his jaw or gums. Liam went with him and afterwards reported that physically, all was well, there was now however, a domestic issue of major proportions.

Apparently, Bert's wife had been on at him for years to get a new set of dentures, in a vain effort to make him look more attractive. He had finally capitulated and this trip was the first time he had worn them, having taken delivery, so to speak, the day before coming offshore.

Sadly, it has to be said, that as far as we were concerned, his wife's primary reason for the new purchase had failed; we still thought him to be his usual ugly self.

However, true love, as they say, being blind, Bert's wife was delighted with her new look husband and, in an effort to re-kindle some passion in her newly toothed soul mate, had booked, at considerable expense, a Dinner Dance for New Year's Eve.

Obviously, explained Bert, this required the purchase of a posh frock, matching accessories, a hair-do with coloured highlights and a baby sitter. All at great cost and effort on her part.

As Liam rather unkindly commented, 'This was the beginning of a new romantic interlude based on a set of teeth.'

The thought of jeopardising these touching domestic arrangements meant that Bert was terrified of breaking the news to his wife. Not only did he have no new teeth, but he had also thrown away his old set in a moment of euphoria and better bite capability.

Fearing for his life, Bert then pleaded with Liam to do him two great favours. The first was to contact the dentist and see if he could obtain a duplicate set within the next couple of days.

The second, and this is where Liam began to display a marked degree of reluctance, was to contact Bert's wife and to break the news gently that her loved one's ability to smile, whilst in mid tango, may have been severely impaired.

Bert reckoned that Liam should emphasise how lucky her husband was to be alive and that only an act of considerable bravery on the part of numerous rescuers had saved him from certain death, etc, etc. His forlorn hope was that the subtle approach would reduce his wife to tears of relief that her newly refurbished husband was still in one piece and the problem of missing teeth would pale into insignificance by comparison.

Having listened to this rubbish, our considered view was that Bert was living in a dream world and that once she knew her hero was alive, his wife would immediately put missing teeth at the top of her of her 'grounds for divorce' list.

Sid, on hearing about the tactics, said it was a wonderful illustration of the gulf between a husband's belief that that he was seen by his wife as a sort of James Bond figure and her long held conclusion that in reality, she had mistakenly married an idiot.

Now, as the need to communicate with the main players onshore was of paramount importance, we approached Ken in the Radio Room and put the problem to him. Realising the seriousness of the situation, when he saw Bert looking like a wizened prune, he said it was clearly a medical emergency and would need to 'pull some strings' with Wick Radio.

We were impressed both with Ken's diagnostic skills and his ability to lie with such consummate ease. Onshore contact was duly made and Liam reluctantly picked up the phone to make the necessary calls. At this point, an anticipatory hush descended on the four of us, as we crowded around the radio—there to assist, of course.

The first call was to the dentist, who having had the facts explained (and embellished) by Liam, came up with some good news, he still had a temporary set which were just as good and would do until a new set could be made.

However, he then frightened Bert to death by saying he was going on his Christmas holidays in the morning, but being an obliging sort, said he would contact Bert's wife and ask her to pop down and collect them.

On hearing this, there was a panic-stricken cry from Bert, as the last thing he wanted was the dentist talking to his wife, especially before Liam had softened her up with his carefully rehearsed tale of bravery in adversity.

Liam, also thinking quickly, said 'Thanks, but don't bother, we'll let Bert's wife know about the problem and arrange for her to come and collect them.'

The next call had us all in a state of trepidation, as Ken asked Wick to contact Bert's wife.

Liam started off by introducing himself, telling her there had been a minor accident, not to worry, how brave her husband had been and that there was just one small problem. We all started to breathe a sigh of relief, Liam had done it - all would be well.

Until that is, Bert's wife, ignoring the bravery rubbish, homed in like a guided missile and said, 'Okay, what's the problem?' Liam took a deep breath and explained about the missing teeth.

Unfortunately, that was as far as he managed to get. She gave a sort of strangled cry, said she wasn't a bit surprised and started to list Bert's many failings. Much of it consisted of his incompetence around the house, the ruined holiday in Rhyl (it rained), his inability to assemble the children's bunk beds and so on and so on. All her radio discipline regarding 'over' had been completely forgotten, so that Liam literally couldn't get a word in edgewise.

Pausing only to draw breath, she ploughed on with the key issue. What about the dinner dance, how could she face her friends with a husband looking like a gummy gargoyle?

Until at last the tirade died away and she subsided in exhaustion.

By now, Ken and I were on the floor, Bert was lying across the counter with his head in his hands and Liam, still clutching the phone, was apoplectic. Bert then roused himself, grabbed Liam and said, 'Ask her whether she is going to collect me bloody teeth!'

Liam, who had by then 'lost it,' thrust the phone at Bert and replied 'Ask her your bloody self!'

Bert backed away in alarm and said 'I can't, you've told her that I'm in the Medic's room having my wounds attended to! Just sodding well ask her.'

We were obviously in the presence of a man 'in extremis.'

Liam, staring at Bert with something approaching hatred, made one last effort and managed to ask whether she would collect the teeth as arranged before the voice of Wick Radio informed us that 'time was up' and broke the contact.

To say that Ken and I were now hysterical would be an understatement. Bert accused Liam of failing to carry out a simple task and get confirmation from his wife, Liam called Bert an ungrateful sod and said it was the last time he would try to help anyone ever again.

Overall, it was a fine illustration of the serious and potentially dangerous nature of offshore work and our ability to 'pull together' in the face of adversity. Again, only a strong brew would restore our equilibrium.

On his return after Christmas, we asked Bert how things had gone, but he was somewhat tight lipped on the subject. This caused Liam, who was still in 'lack of gratitude' mode, to say it was a pity he couldn't have been quite so tight lipped when it really mattered. I can't recall Bert's actual reply but it had something to do with Liam's parentage.

During the course of that year the Buoy began to dominate our lives. Mooring lines snapped, hoses were damaged and needed changing, weather was bad and the tanker either couldn't moor or had to pull off. The electronic controls were notoriously fickle and would fail at the most inopportune times.

During the mooring sequence huge 'stoppers' were supposed to swing into the mooring rope reel when the tanker was in position. They would then hold the reel in a stable position and the tanker crew could relax.

The stoppers were pushed in place by huge rams and were supposed to operate automatically when the mooring rope and the hose were in their fully extended position. However, as the system was also unreliable (surprise, surprise,) the tanker had no idea whether the stoppers were safely in position, or not.

This meant that every time there was a tanker to moor or release, we had to go across to ensure the system operated properly and confirm that the stoppers *had* swung into place. Remember all this was before the advent of computerised thingies. We did initially have a signalling system based on 'telemetry' but like semaphore it never really worked in the dark.

Depending on weather, there were several means of transport available, each one more 'hairy' than the other. The best way was by helicopter, the next best was on the back of the supply boat and the worst way was by descending on the end of the crane into an inflatable dinghy, which was normally carried by the stand-by vessel for rescue operations.

Even so, I use the word 'best' in its loosest sense as the latter methods of transfer involved an adventurous leap onto an iron ladder fixed to the outside of the Buoy's fenders. This provided a significant adrenaline rush, whether you wanted it or not, especially if there was the usual sea swell running. We were always taken aback at how different the prevailing sea state looked when observed from the platform, some forty-five feet above the waves, compared to when you were being thrown about in the inflatable dinghy.

This strange new existence was as lunatic as working underground had been, but with the added advantage of fresh air. I remember that Her Majesty the Queen had once referred to 'The excitement and romance' of oil discovery in the North Sea. We certainly had plenty of excitement—but romance?

As the weather became colder and our trips to the Buoy became almost routine, we spent many hours debating whether there was a way in which we could help the tanker to ensure the 'stoppers' were in place. Discussions centred on the need for some sort of signal which would operate consistently and would be visible or audible to the tanker in both day and night conditions.

It was finally agreed that a visible signal was the answer and someone suggested that what we needed were green and red lights to identify the position of the stopper, which could be seen from the ship's bridge.

I remember asking George, our electrical technician whether it would be possible to wire such an arrangement to the existing system on the Buoy.

'I don't see why not' he answered and with that, a unique answer to the problem emerged.

George carried out a survey and pronounced our idea to be feasible. I said, 'Good, you start sorting out what you need and I'll order some lights.'

This rash promise proved to be more of a task than I had envisaged. However, filled with premature enthusiasm, I sent an order to our purchasing department requesting a set of Traffic Lights, as used by road repair contractors. This was initially met with some disbelief and numerous queries on a theme of, 'Were we trying to take the mickey,' and 'Whether we were bloody mad.' In the end I had to go onshore and present our case to various important and somewhat cynical people.

What we had failed to realise was that the people onshore had little or no idea of what the tanker's problem really was. Fortunately, after several meetings and numerous sketches, I was successful. Perhaps this was partly due to my introducing vague threats regarding loss of revenue, at key moments. Don't you just love the way money focuses the mind?

Anyway to their credit, the traffic lights were ordered and duly delivered to us for installation. George and the lads excelled themselves and within the week we had completed and tested the lights. Liam and I flew over to the tanker to witness the final test and to obtain feedback from the Captain and First Officer. We were filled with a certain degree of trepidation as the tanker approached; it was one thing simulating the action of the lights, but this was for real and we were conscious that our credibility was at stake.

In the event, the system worked perfectly and, as the stoppers swung into the locked position, the lights changed from red to green. The Officers were delighted and there were congratulations all round. Our pride knew no bounds, as we went back to the platform to send telexes to everyone onshore.

In the meantime, Liam kept going on about tangible rewards, until Sid reminded him that he aught to be bloody thankful the crazy system had worked. We were also grateful for the trust the Purchasing Manager had shown in us, even though he couldn't really visualise what it was we wanted to do with the lights.

Inevitably, word spread of our novel solution and there was an extensive article in the *Manchester Evening News* entitled 'It's Green for Go', with a sub-heading saying,

'Life on a North Sea Oil Rig is dangerous, cold and hard work. Every little luxury has to be flown in. But what did they want with a set of traffic lights?'

Fame at last and a better press than the *Times* Insight article. One up to us in the newspaper war—I think! We could even forgive them for calling us a 'Rig.'

It was the start of the final stage of the Monte Carlo Rally, we were lying second behind Stig Blomqvist in the Saab 96 and Roger Clark was less than a minute behind us in the Ford Escort RS. Suddenly my co-driver began to shake my shoulder violently, 'What the hell are you playing at?' I cried, as he continued to shake.

I opened my eyes to see the Ken bending over me, shouting, 'Wake up you lazy sod.' I realised I had been fast asleep, it was dark, I wanted to strangle him and I was sure I could have beaten Blomqvist.

Having stopped shaking and shouting, he then gave me the good news. Apparently he had just received a call from the tanker to ask us to stop pumping oil.

This little episode started at 0430 hrs on Christmas Day. 'The First Officer wants to talk to you and it sounds urgent,' said Ken. I got dressed and made my bleary way to the radio room and made contact with the tanker. The First Officer apologised and gave me the bad news. The weather was atrocious and they were afraid the ship would be driven into the Buoy by the surging waves.

As the sea state was worsening rapidly, he had made an initial decision to tell us to stop pumping. But as conditions were worsening and time was against them, he had gone into non-technical mode and taken an axe to the mooring rope. However, as it severed, the ship moved rapidly astern and unfortunately the oil delivery hose, which was still coupled to the tanker, snapped under the strain. Having accomplished what was to them, a nautical manoeuvre, they were now lying off, about two miles away and steaming safely into the wind.

Having dried my tears, I realised we would have to go over to the Buoy and estimate the damage. This, of course, meant I would have to call out the troops, because if there's one thing you learn as a coward, it is that you don't go into the darkness alone.

This gathering of the clan took some time and a good deal of swearing, but after some dissent and much grumbling, our 'expert team' consisting, as usual, of Liam, Bert and me, was ready to go.

Allan, our helicopter pilot, was roused and with great reluctance (this is an understatement), had his mechanic prepare the chopper for the flight. It was now 0600 hours (or 6am for Radio 2 listeners) and at first light, we set off on the short trip to the Buoy. It was a bleak, miserable morning and the water looked like the bottom of a saucepan. The wind had diminished a little, but the sea was still running in long high swells. Allan took us around in a long circle to come in against the wind and we peered out of the windows to see the extent of the damage

Did I mention earlier that the Buoy was painted in a tasteful shade of bright yellow? Well that was yesterday. Today, in the growing light we could see that most of the structure was now a sort of nasty brown colour with the odd yellow streak showing through. I asked Allan to hover on the hose side so we could see the result of the tanker's emergency action. The cut mooring rope was lying downward from the reel into the water, where it was wrapped partway around the fenders. The hose occupied a similar position only with its end hanging in the water.

Well, well I thought, nice one tanker and a Merry Christmas to us. (That's not exactly what I thought, but this is a family book.)

Allan circled around once more and then set us down on the helideck. We clambered off and he disappeared back to the platform.

Where to start? Even Liam was silent for a few minutes but, as we couldn't go back, we knew something had to be done. Being loyal employees, we were aware of the loss of revenue (the Chancellor expects, etc.) and the need to make the Buoy ready for the tanker to hook up again, as soon as possible.

Gingerly, we set off down the steps from the helideck, amid growing evidence that although the platform had stopped pumping prior to the tanker's release, the hose had still been full of oil. Consequently, when it snapped it had flailed backwards and spread the contents liberally all over the structure. There was oil everywhere, dripping down the steelwork, covering the steps and handrails and now methodically covering us as we made our way about.

A word of explanation here; do not imagine the oil I am talking about bears any resemblance to the oil you buy in a nice shiny can, which flows like liquid gold when poured. Our oil was thick, dark brown, smelled awful and when exposed to cold air, started to congeal, with a consistency roughly similar to that of treacle.

It also had an unerring ability to stick to skin, clothing and all metal surfaces. The best analogy for older readers is to imagine opening a tin of under-seal and smearing it all over yourself.

(For younger readers, under-seal was a black sticky mess similar to axle grease. You had to spread it on the underside of your new car in order to prevent it rotting away within the first six months. For younger readers, axle grease was—oh just look it up yourselves on the interweb thing).

The other thing that is etched on my memory was just how miserably cold it was. The wind was still strong and where we were working lacked any form of shelter. After about an hour, even our insulated overalls were no proof against the biting wind.

We were cheered however, by the sight of our old friend the supply boat, who was circling around at a safe distance and seeing three bedraggled figures on the Buoy, made his way to us. He proved to be of invaluable assistance, as with his help we dragged the mooring rope and hose back onto the Buoy. Fortunately the supply boat carried spare lines and lengths of hose, so we were able to rectify the damage and make good the ruined sections of hose.

What we couldn't do, was get rid of the oil, which was slowly solidifying on walkways, handrails, stairs, trunking and our overalls. Again, the supply boat came to the rescue and in a remarkable feat of seamanship, came alongside and threw his water hose across the gap. At a given signal, he switched on his pump and we hosed the Buoy down as best we could with his store of potable water.

Unfortunately the spray demonstrated an unerring ability to follow us wherever we went, so we were soaked by a mixture of cold water and congealing oil—just the thing to enhance our already debilitated state

It was now about twelve o'clock and as our overalls were slowly solidifying, we felt that a brew was called for. Slipping and sliding our way up the iron stairs, we fell into the small cabin and over a steaming cup of Darjeeling's finest, began to question very seriously just who it was had told Her Majesty about the 'romance' of offshore life—the bloody idiot.

Whilst deliberating on this calumny, we became aware of a small but increasingly important niggle. Back on the platform, a sumptuous Christmas dinner was being prepared, while we were slowly congealing in a freezing cold steel hut out of reach. There would be crackers, ill fitting paper hats, useful little tools, pathetic jokes and pudding with artificial brandy sauce, all washed down with vintage orange juice.

And horror of horrors, we might miss it all.

However, as we were rough tough oilmen we dried our tears, staggered to our feet and finished the job so the Buoy was ready to couple up to the tanker, as soon as the weather abated.

At last it was time to make contact with the living, so I called Allan on the radio and said we were ready to be taken off. Unbelievably, as far as we were concerned, his reply was not what we wanted to hear.

He said the Buoy was heaving too much for him to be able to land on the deck and we would have to wait until the sea calmed down. This news went down rather badly with the stalwarts and our intrepid flyer was subjected to a rather jaundiced analysis of his capabilities, parentage and cowardice in the face of adversity. After this, we felt better and made another brew—having estimated there was still time to get off the blasted thing, as our Christmas feast was scheduled for 3 pm.

After about ten minutes, I was persuaded to call Allan again, but to no avail, the motion of the Buoy was still too extreme and Allan suggested we hang on for another hour and review the situation again. We settled down again on the bench and Liam, in his own inimitable style, reassured us by continually estimating the time needed to transport us, clean up and still be in time for dinner. Thoughts of food were also starting to dominate, as we had nothing to eat on the Buoy. Having violently disagreed with each other's choice of the perfect meal, conversation was flagging when Liam dropped another of his bombshells:

'What about Jimmy Young?'

We stared at one another, 'Oh bloody hell, I forgot about him,' I said. 'What time is it now?'

A word of explanation is needed here. As the offshore industry was a new phenomenon and no one was trying to kill President Ford, there was a good deal of Media interest in the plight of our brave lads roughing it in the middle of the North Sea.

Our Company had been contacted by the BBC, who thought it would be a great wheeze if Radio Two's Jimmy Young could talk to one or two of us during his live Christmas Day broadcast.

Sid and I had been fingered for interview by the great man and the Head of Public Affairs had stressed in no uncertain terms, 'That we were to be polite and vague, as he didn't want a repeat of the Sunday Times fiasco broadcast on National Radio.'

Now, for readers under the age of 70, I should give you a potted biography of the said J.Y, as he liked to be called. He started out as a singer with one of those wobbly voices, who were very popular at the time. He had a great many hits, most notably the first, *Unchained Melody*.

His second hit was the title song from the film *The Man from Laramie,* which clinched his position as the UK's second biggest selling artist of the year. In 1960, J.Y. introduced the radio show *Housewives' Choice* and began a new career which lasted more than 30 years. He then hosted his own morning radio show, mixing records with consumer information, discussions on current affairs, and interviews with figures in the public eye, including several Prime Ministers.

So you can see an interview with Jimmy Young was not something our Company would want to miss. I was similarly impressed; in my opinion, if being interviewed by J.Y. was good enough for Harold Wilson, it was good enough for me.

Liam consulted his watch and said, as if it was my fault, 'You're too sodding late, it's nearly two o'clock.'

At a stroke, my dream of being asked to take over from Robin Day began to fade and recognising there would be no offers from a grateful BBC—I needed someone to blame. This was a wonderful catharsis, as I mentally listed the First Officer, Allan, the weather (always a good bet) and fate.

As my aspirations turned to ashes, our interest returned to getting off the blasted Buoy in time for our Christmas dinner. I called Allan again and said, 'It's a miracle, the water resembles a mill pond, we're hardly moving, so could you come over for us straight away?'

Allan, contrary to our wishes, said he still felt the conditions were unfavourable, but asked us to give him the Pitch and Roll readings of the Buoy.

These readings were taken from two gauges mounted on the highest level just under the helideck. They were crude affairs and as an accurate measure of relative movement, our pitch and roll gauges were just about useless.

They confirmed our worst fears; both pointers were performing a sort of mad tarantella and reading in excess of five degrees. Now I realise that five degrees doesn't sound like much, but remember, as we were over forty feet above the water, this meant that the helideck was moving a considerable distance from side to side. So, contacting Allan again, I said the readings were between two and three degrees for both pitch and roll.

There was a momentary silence and then he replied, 'You're a lying sod, I've got my glasses on you and even from this distance I can see you're bobbing about like a fishing float.'

This news didn't go down too well with the lads, who began to plan a suitably lingering death for Allan.

It was now about two-thirty, the prospect of our Christmas dinner was rapidly diminishing and I didn't want to break radio contact.

However, after a minute or two of radio silence, Allan relented and said 'Ok, I'll come across but there's no way I am going to risk setting the helicopter down on the deck. What you'll have to do is crouch on the side of the helideck, facing the wind. I'll come in and hover at about six feet and as soon as I wave, you run across and fling yourselves in. Have you got that?'

We looked at one another, dismissed all worries about safety in favour of dinner, quickly agreed and gathering up our overalls took up our positions. It was then that the true extent of the Buoy's movement became apparent. The helideck was shuddering from the force of the waves so, as there were no railings around the perimeter, we hung on like oily limpets.

Thankfully Allan was as good as his word. We watched in trepidation as the wind buffeted the approaching Bolkow. Allan slowly manoeuvred until he was hovering above the centre of the helideck and on his signal we dashed across and dived through the open door.

The moment we landed aboard in a jumbled heap, he took off in a steeply angled climb away from the Buoy and five minutes later we were back safely on the platform.

I then had the longest, hottest shower I have ever taken before or since.

Thanks to Allan, we were just in time for our dinner—though we were somewhat miffed to find that no one had missed us.

Allan was now, in our opinion, the best Helicopter Pilot in the world and woe betide anyone who dared to criticise him. He was in fact, a superb pilot and what he did to get us off the Buoy required both nerve and considerable flying skill.

As time went on, we had a number of other interesting flying experiences with him, as we shall see.

Speaking to Sid over dinner (hats on, etc.) I lamented the fact that the BBC had been unable to broadcast my valued opinions on life and other geopolitical issues. He then confessed that his conversation with Jimmy Young hadn't exactly gone to plan. Apparently, reception was very poor (curvature of the earth again?) and J.Y. only had time to ask a couple of penetrating questions like, 'I believe you are called the OIM, could you tell us what that stands for?' and, 'How do you feel about spending Christmas in the middle of the sea?' before communication failed altogether and that was it, no further contact and no chance for Sid to give the listening public in-depth answers.

I said to Sid, 'It's probably a good job you didn't get a chance to answer that last question or the head of Public Affairs would have ended up apologising to the BBC again.' Sid just snorted, but didn't deny the very real possibility.

He then dropped the biggest bombshell of all by telling me that while we felt privileged to have Jimmy Young, our biggest rivals in the industry had managed to fly Shirley Bassey out to one of their platforms. Imagine that voice, those dresses, those curves—no, well perhaps not.

It's the kind of major disappointment that can make or break a person's character. However, the orange juice was flowing freely and there was the promise of a film to come. We were obviously capable of rising above such an emotional shock as the loss of Shirley Bassey, although it was touch and go for a while.

As we spent a good deal of our time flying back and forth to the Buoy, we got to know Allan pretty well. He was an ex-Army helicopter pilot and had seen considerable service overseas. He confessed to a deep abiding love (lust?) for Agnetha Falstog, the blonde singer from Abba, but hastened to add that what really attracted him was the band's music; the fact that she habitually dressed in Lycra and thigh length gold boots was purely incidental. Yeah, right.

As I said, Allan could make the Bolkow do extraordinary manoeuvres and often, as we returned from the Buoy, Liam would ask him if we could have a bit of fun on the ride home. Although I suppose (oh alright - know) he shouldn't have, Allan could never resist the temptation to show what the Bolkow could do. Consequently, we had many exhilarating moments going vertically upwards, zooming sideways and diving at the water before levelling out just above the waves.

On one occasion, Bert, Liam and I had been over to the Buoy to accompany a technical specialist who was to install some sophisticated wave recording equipment. He needn't have bothered. As Bert said, looking daggers at Liam, 'We already know all there is to know about waves.'

Anyway, job done, we called up Allan and boarded the Bolkow. Liam then uttered the fatal words, 'Hi Allan, give us a nice run back.'

As far as Allan was concerned, this was tantamount to receiving flying orders. He grinned and shouted 'Right, tighten your seat belts!' We buckled up, having learned the hard way that this instruction was rather important, and away we went.

It isn't possible to recount in detail the sequence of events that followed, but we certainly carried out a number of our favourite routines. The first of which was to come off the helideck and take a steep descent to level off about six feet above the water. We then headed straight for the supply boat and just as it seemed certain we would crash into his bridge, Allan pulled back on the controls and we zoomed over his antenna. Fortunately, this didn't bother the Skipper as he was well used to Allan's rather unique flying style and anyway, we always gave him a sort of cheerful wave as we shot past. On gaining some height and fast approaching the platform, we would track around the structure in an ever-tightening arc until we were flying on our side, parallel with the water once more, with G-forces keeping us firmly in our seats. Finally levelling off, Allan's favourite *piece de resistance* was to fly just above the sea once more and hurtle straight towards the platform. Again, just as it seemed we would fly into the legs, he would pull back on the controls, lose speed, level off and land gently on to the helideck. Wonderful!

People talk these days about the knuckle-biting rides in places like Alton Towers, but for me, they pale into insignificance when compared to the adrenalin rush we got from Allan's helicopter aerobatics. We were aware that the experience was unique and due entirely to the skill of the pilot, as opposed to being thrown around on a predictably safe set of rails.

We loved every minute of it and as we came to rest we would let out rather undignified whoops of joy and slap Allan on the back, gently of course. Then, despite feeling rather dizzy, we would climb out and make our way off the helideck.

However, as we decanted on this particular occasion, we became dimly aware of a rather different noise coming from the back seat. With mounting horror, we realised we had completely forgotten our specialist.

He obviously thought he was in for a straightforward five-minute flight from the Buoy to the platform. Instead, he had been treated, without warning, to a series of stomach-churning aerobatics that had terrified him.

As we watched, he glared at Allan with eyes like organ stops and shouted, 'You're a bloody madman, what the bloody hell are you playing at? Get me out of here!'

Allan, who, even though he still had his headset on, was aware that his name was being taken in vain, didn't help matters by asking Bert, 'What's the matter with *him?*'

At this, the specialist nearly had a fit and renewed his tirade against Allan, our Company, the Government and sundry other entities. It began to dawn on us that we would have to calm him down and retrieve the situation, as the last thing we wanted was to get Allan into trouble.

Liam took control and guided the specialist off the helideck and into my office where we sat him down with a cup of strong tea. I couldn't for the life of me figure out how we were going to get out of this dilemma, until Liam, going into full inspiration mode, realised the guy was still traumatised and incapable of logical thought.

Sitting down next to him, he gently began 'Sorry about that, we should have told you earlier.'

'Told me what?' demanded our victim between gulps of tea.

'Well, whenever this type of helicopter reaches a given number of flying hours, the pilot is required by the CAA to perform a number of specific test manoeuvres. What you were involved in was just such a test. Obviously Allan had briefed us beforehand, but we must apologise for forgetting to let you know.'

Bert and I stared at Liam and then at the specialist. Would he swallow this highly imaginative garbage, or would he respond with something technically pertinent like 'Don't talk bloody rubbish?'

Our friend looked at Liam and asked, 'But why would he do such a test with passengers on board?'

I had also been asking myself this question, but Liam had the answer.

'It's all to do with balance. The test requires the helicopter to carry a minimum load for the results to be effective, and we had volunteered to act as ballast.'

At this, the specialist gave an exhausted sigh, shook his head, finished his tea and said he was going to bed.

I turned to Liam. 'Do you think he believed you?'

'Let's just say we've sown the seeds of uncertainty. If he starts to ask questions about this onshore, he could look like a prat. After all, there's nothing like a bit of healthy terror to focus the mind.'

Although, once again, I had no idea what on earth Liam was talking about, I was happy to be reassured; and as it turned out, our friend went ashore the following morning and we heard no more about the episode.

Speaking of tests, I was in my office one evening when Allan put his head around the door. 'Are you busy?'

'Not really, what's the problem?'

'No problem, Colin, my mechanic, has just finished a service on the machine and I now have to do a stall test. I wondered whether you would be interested in coming along.'

'Definitely,' I replied, not having the remotest idea what this entailed.

So, a few minutes later Allan, Colin and I were all strapped in and ready for the off. It was a pleasant summer evening, just perfect for a nice ride.

As we took off, I asked as casually as I could, 'What's a stall test?'

'Well it's nothing really; I have to go up to a reasonable height and simulate a stall. To do this, I switch off the engine and we drop out of the sky while Colin is making observations. We'll pull negative gravity for a short time until I switch the engine on again and we regain control of the helicopter.'

'Oh, right.'

This was one of those 'mixed feelings' moments. On the one hand, I was sorry I had asked and on the other, I rather wished I had said I was too busy to come on the trip. But, being a coward, I said the only thing I could think of in the circumstances.

'Sounds great.'

And indeed it was, although I have to say when Allan switched off the power, it did cross my mind that I should have told Allan to take Liam. Still, where else could you get paid for buzzing around the sky, pretending to be an astronaut in the capable hands of a master flier.

A couple of years later when I was flying in a crew-change Sikorsky out to a more northerly location, there was a call from the flight deck for Mr Page to go to the front. Greatly puzzled and with some concern, I did so and there, covered in gold braid, was Allan in the driving seat.

Looking to a more stable future, he had joined British Airways and had recognised me from the flight list. I sat with him for the duration of the flight, but it didn't seem to be the right time or place to ask about the possibility of aerobatics, just for old time's sake.

I did mention that his new job entailed a rather more formal approach to flying than when we worked together, to which he replied, rather sadly, 'I suppose we had to grow up sometime.' But still, it was great to see him again.

13

The Pressure of Work or, a Problem Shared

In between our flying activities, we also found time to resolve a number of highly technical maintenance problems and you may be surprised to learn that Liam and Bert figured large in their resolution. You may also think this can't be correct, as there must have been many others involved. Well, yes, there were one or two other people carrying out essential maintenance activities, although at the time we had very few staff. It is true to say however, that most of the chaotic happenings involved the three Musketeers, with me in nominal control.

For instance, a message arrived one day from the Production Supervisor. His men had been hosing down the wellhead module and the open drains were no longer taking away the water. Could we please solve the problem?

Now although these weren't his exact words, the gist was clear; get it fixed forthwith. So we went down to the wellhead module to take what is known as a 'professional look'. The wellhead was the area where oil at considerable pressure arrived from the wells before being piped into a series of process vessels, something like those on the back of a petrol tanker, only about 10 times the size.

We lifted various access grids, but couldn't see where the problem lay, so, realising we had a serious blockage somewhere in the system (not called engineers for nothing) we sent for the local Drain-Rod men.

Rising to the challenge and being blissfully unaware that our drains were somewhat different from the household variety (probably because we were careful not to tell them) they and their kit arrived on board.

Being young and foolishly optimistic, they set to work on the drains, assisted by numerous drain rods and a powerful water pump. Soon there was water everywhere except, unfortunately, in the drains. After about two hours; during the course of which, rods were bent, we were soaked and their pump began to smoke and trip out on overload; we called a halt. This was in order, as Liam said, to review the situation and find his wellies.

The Drain-Rod operators were, by this time, becoming somewhat concerned at the lack of progress and, worried about their reputation, decided to send for a more powerful pump. This, they assured us, would sort it, as it had a delivery pressure capable of blowing up the average house. We were suitably impressed and once the fearsome machine was assembled, we donned our wellies (chosen by the Company in a tasteful brown colour with steel toe-caps) and the lads set off again to blast their way through the blockage. True to their word, the new pump took no prisoners; it was like attacking the drains like an Apollo space rocket with a sound similar to that of a 1000cc Yamaha.

Imagine then, our surprise and delight when great fountains of grey water, accompanied by lumps of drilling cement, started to erupt from various grids and manholes. It was like watching Vesuvius, but without the heat. Cheers all round, but like lots of things offshore—somewhat prematurely.

Now I may have mentioned earlier that drilling people couldn't get to grips with the fact that the platform was not just a bigger than average drilling rig. The idea of sharing the facilities seemed to be intellectually beyond them and unknown to us they had been happily cleaning out their cementation system and washing the slurry into our open drains. (Don't ask me why they use cement; this book isn't about the mad world of drilling).

Unfortunately though, as our drains were designed to take the occasional flow of reasonably clean water, the massive amounts of liquid cement simply started to solidify along the pipes and in particular at bends.

Obviously, in order to apportion blame, we needed to have a word with the Toolpusher, but the thought of trying to explain the error of his ways to the same lunatic who had tried to kill me earlier didn't appeal to any great extent. Trying gamely to delegate, I suggested that Liam or Bert might like to 'have a word', but to my surprise, they both declined. There was only one other option; Sid would have to reason with him. After all, that's what bosses are for.

We convened a meeting in Sid's office and, to add verisimilitude to our contention that the drilling department were culpable, we placed several lumps of cement on Sid's desk. Our psychopath stared at the 'concrete evidence' for a minute or two and then said, 'Is chiment, what you got it der for?'

Bert, who was still dripping slightly, glared at him, 'What the hell do you mean, "What we got it der for?" It's what we've just dug out of our pipes after your bloody daft lot have flushed your horrible muck into our drains.' Bert's command of the language was quite lyrical when he was riled. However, leaving aside our madman's patois, his idea of an abject apology left a bit to be desired.

The gist of it was 'What did we expect?' Apparently he had drilled for oil all around the world and never had any problems with *drains.*

'Because,' he said, 'On my drilling rigs, we simply washed any cement and other waste products straight overboard, so we never needed drains.'

After this lesson in twisted logic, he put the piece of cement down and left the office.

I looked at Sid, who shrugged and said, 'I guess the message is clear, the daft sod has no use for sophisticated drainage systems.'

As an answer to our problem, this statement fell someway short of satisfactory and I could see that in Sid's eyes, the problem was ours for installing silly little drains and it was up to us to sort it out.

To cut a long story short, we thanked the Drain-Rod lads for their help, reassured them that it wasn't their fault and shipped them ashore.

However, as we now faced a future without drains, there was no option but to scaffold under the platform in order to find and remove all the solidified bends and tee pieces. Replacement sections then had to be fabricated onshore, shipped out and fitted, after which the scaffolding had to be removed. We won't go into minor details like cost; it will only make you sad.

Have you ever wondered how you build scaffold on the underside of a platform to which there is no access and the North Sea is going berserk some fifty feet below?

The answer of course is mind-bendingly simple. All you need is a ten-foot scaffold pole, a two-foot scaffold pole and a scaffolder who knows no fear. You clamp the two-foot piece to the bottom of the large pole so that it forms a cross. You then lower this assembly down through a hole in the deck and attach the top of the pole to a girder on the underside of the walkway. You now take the fearless scaffolder who lowers himself through the hole and, in the manner of a fireman slides down the pole until his feet land firmly on the cross-piece.

Other members of the team then pass more poles down to our man, who attaches them to his pole until he has the makings of a framework. He is then joined by more maniacs who, within a short space of time, have created a walkway. See - I told you it was simple.

Oh and just to make things more interesting, this little exercise took place in November, a lovely month for crawling about on bendy planks under the platform. All of which made it increasingly difficult to cement relationships with our drilling colleagues (Joke? I think not.)

We had another adventure which, this time, was associated with the removal of gas from the oil as it came up from the well. This task formed part of the production process and was done by dropping the oil pressure in large vessels, so that the gas bubbled out of the oil. The (at the time) unwanted gas was then piped to the flare boom and burned off. The remaining oil was taken from the pressure vessels and pumped through a pipe on the seabed over to the Buoy and onto the waiting ship. As you can imagine, leaving gas in the oil could have a somewhat detrimental effect on the tanker as it would leak out of the now depressurised oil and expand, so that the ship was in danger of becoming a balloon.

There was an awful lot of gas and the only way for us to dispose of it safely was to set light to it. Gas flaring was the recognised method of disposal in the early days and it took an edict from Tony Benn, who was the Minster of Energy at the time, to make the oil companies either re-inject the gas back into the reservoir, or pump it ashore for sale. This also provided a nice little earner for Tony and it meant that the Government could raise lots of lovely tax to spend on something worthwhile - like British Leyland.

Anyway, at the time we were flaring millions of cubic feet of gas every hour; the trick was how to ignite it as it came out of the end of the flare boom at something like warp speed.

The system we had installed was a giant version of the type of igniter you would use to light a gas cooker at home. You pressed a button at the base of the flare and a piezo-electric spark was emitted at the tip and, 'hey presto', the gas went 'whoof' and a massive flame incinerated anything silly enough to get in its way.

In theory, that is.

In reality it didn't; the gas steadfastly refused to ignite and we simply blew huge clouds of noxious gas into the atmosphere. This was okay up to a point; but if the wind blew the gas towards the platform instead of away from it, we then had what might be called a problem involving mass poisoning.

Over the next couple of days (and nights) we wrestled with the ignition system, but to no avail; the damned gas refused to ignite. We then started to seriously consider trying another of Liam's 'good ideas.' It shows just how desperate we were that we gave him the go ahead to fabricate and install his latest brainchild.

In essence, his solution was probably the least sophisticated idea he had yet come up with. First, he removed two pulleys from some lifting gear and fixed one at the base of the flare and the other at the tip some 60 feet away. He then wound an endless thin wire rope around the pulleys and, as he had fixed a handle to the pulley at the base, he could wind the rope from the bottom to the top. Genius or what?

The next bit was the highly technical part—an oily rag was tied to the rope, the gas valve was opened, the oily rag was ignited, the handle was wound frantically, the oily rag reached the flaring gas, the gas blew the flame out and the rag disappeared into the sky.

One small setback for mankind—and Liam.

However, nothing daunted and having a plentiful supply of oily rags, we tried again, and again, and again. By now, a considerable amount of gas was filling the night sky and a large number of blazing oily rags were floating gracefully into the sea.

Worse still, we were beginning to look at Liam in a new light (but obviously not from the flare)—that of the bungling incompetent kind. There was also a growing awareness by the participants of the need to distance themselves as quickly as possible, from any further association with this idiotic enterprise.

In the end, Liam was tranquillised, the gas was shut down and we repaired to the galley for mugs of tea and a chance to review the situation.

Nowadays there are lunatics called Management Consultants who, when providing what is laughingly known as 'Business Improvement Training', insist on replacing the word 'problem' with 'opportunity'. This is supposed to wrap you in a warm, positive thinking blanket and hence provide mental stimulus to solving the problem, sorry, I mean the opportunity.

Can you imagine Jim Lovell uttering the immortal words 'Houston, we have an opportunity?' I don't think so.

All I can say is that on this particular evening, it was a good job we hadn't yet had the benefit of such radical thinking. As far as we were concerned, we had a formidable 'problem' and a solution had to be found.

The solution, when it emerged, was simplicity itself—or so we thought. We believed we could ignite the flare by firing a Very pistol into the gas cloud and, with a satisfactory bang and a whoosh, or so the theory went, the flare would be lit. Even Liam thought it might work, but this did little to bolster our confidence. One minor problem was that the Very pistol we had on board fell within the definition of a 'firearm', which meant it was locked in a safe in Sid's office.

What's more, we had to convince him that there was no other viable option and, as he was the custodian of the pistol, he could have the honour of firing the cartridge. I don't recall Sid's actual words, but phrases like 'You're all bloody mad!' were much in evidence. However, curiosity got the better of him and we all trooped back to the base of the flare boom. After some persuasion, the operator reluctantly opened the gas valve and, taking careful aim, Sid pointed the pistol skywards and fired into the gas cloud. There was a satisfying 'whoosh' and a streak of flame as the cartridge sped away. Unfortunately, that was the only flame we saw.

Once again, the gas had beaten all attempts to let us say those immortal words 'we have ignition'.

We tried, unsuccessfully, several more times, but by then Sid was beginning to regret his offer and was muttering about not having any ammunition should the pistol be needed for rescue purposes. Bert convinced him to keep trying by stating the obvious fact that if we continued to vent gas to this extent, we would all be asphyxiated anyway and he thought that could come under the heading of an emergency. In the meantime, while Bert was working his magic on Sid, it dawned on us that if we continued to fire into the centre of the gas, its sheer volume and density would ensure there was insufficient oxygen to provide an adequate gas/oxygen mix for ignition. (As in your basic physics, we weren't called engineers for nothing)

Convincing ourselves that this was the answer, Sid reloaded once more and this time he fired at the edge of the gas cloud. There was a moment's hesitation and then, with a blinding flash of light, the gas burst alight.

Being British we danced a dignified little jig and felt, quite literally, the warm glow of success, accompanied by singed eyebrows and Sid taking all the credit.

From then on the Very pistol became the standard means by which we lit the flare. Liam and the electrician quietly ignored the problem with the Piezo electric system and the escapade with the oily rags became folklore.

I was recounting the Very pistol episode to Allan our pilot one evening and he told me a lovely tale about an incident which occurred during the Second World War, involving its use by the Royal Air Force in a tropical posting.

It seemed that their communications system was very inadequate, so the Station Commander decided the signal for the Squadron to 'scramble' was for the Control Tower personnel to fire a Very pistol. One morning, whoever was to fire the pistol was so excited by an impending raid he forgot he was inside the tower when he fired. Apparently it was pretty exciting as they tried to alert the aircrew, while at the same time dodging the flare which was flying around the room, emitting showers of sparks.

Fortunately, despite the fact that the pilots were a bit late in engaging the enemy, the only casualty among the ground crew was a chap standing under a palm tree being hit by a coconut shaken loose by a bomb blast.

A year or so later, I was on a platform further north and having just finished a prolonged shutdown, we were ready to light the flare. The difference in this case was that the flare was remote from our platform, as its purpose was to take gas from several nearby installations. The flare was tethered to the seabed and the top section stuck out of the sea like a giant firework about seventy feet high. As, once again, we couldn't light the thing by the designed method (there's a surprise,) the Very pistol was retrieved from the OIM's safe. Now, being remote, we obviously couldn't light it ourselves, so we asked the standby boat to do it on our behalf.

Sadly, it was a case of history repeating itself and after numerous attempts; they also failed to light up the North Sea.

Enter the Royal Navy.

The reason for their presence was to keep an eye on the Russian 'fishing vessels' sailing in close proximity to our installations. 'Fishing' in this context bore little resemblance to reality as these trawlers were brimming with radio antennae, radar scanners, underwater listening devices and lots of crew-members with powerful binoculars. The final clue to their real purpose—and this we felt to be a fundamental weakness in disguise on the part of the Soviet Empire—was the complete absence of fishing nets. A small but important mistake I think you will agree and one that John le Carre would never make.

Buoyed with the excitement of being involved in the 'cold war' with some real Russkies (as we in the maritime service called them), we used to try to make radio contact with them. This cunning ploy consisted of asking subtle questions like, 'Hello Ivan, how goes the fishing?' or 'Caught much today?'

To our surprise, we never received any answers and they just continued circling. Perhaps it was our pathetic attempts at humour that baffled them. On the other hand it could have been because our Radio Operator spoke with a Geordie accent.

Invitations to come on board and have a look around were also ignored. We felt this was a missed opportunity on the part of our sinister visitors. 'After all,' as Sid remarked, 'Detailed drawings of our platform were lying around all over the place.'

Liam felt that their obtaining such documents could be a political coup for us, as he was convinced that most of them bore no resemblance to what we were actually building.

Anyway, on this particular occasion, it turned out that a Royal Navy Corvette had popped round to reassure us we were in safe hands and had been watching the valiant, but failed, attempts by the standby boat to light the flare. The Captain, in a fit of *esprit de corps* and poor judgement, made contact with us and offered his services, saying that his 'lads' would welcome the break in monotony. An offer we gratefully accepted, suspecting there was fun to be had.

He then outlined his plan, which was to sail close to the flare then, firing their Very pistol with unerring accuracy, ignite the gas. There also seemed to be a clear implication in his offer, that this would be a simple task for the Royal Navy. We agreed that only the Senior Service could come up with a plan of such devastating simplicity and waited with growing interest—if only they knew what we knew.

After a short briefing from ourselves regarding our need to ignite the gas as soon as possible, the corvette, in a display of impressive seamanship, turned about and steadily approached the flare. They then slowed down at just the right time, a manoeuvre obviously designed to give their pistol aimer a superb shot at the gas cloud. Unfortunately, they then committed a cardinal error, probably, it must be said, because we had somehow failed to mention it during our briefing; the seaman, having taken careful aim, fired the charge straight and true into the centre of the gas cloud, which immediately doused the flame.

The Captain called us. 'Sorry about that,' he said in a light-hearted voice, 'My chap must have just been a bit off centre due to the prevailing wind. Not to worry, we'll just go around once more and get it right this time.'

By now, all work on the platform had stopped and people were lining the rails to see the fun. We watched as the ship approached the flare and you couldn't fault the skill of the crew, they brought the vessel in slowly and once again, at just the right time they fired off another flare.

As before, they fell into the same trap; the flare went straight into the centre of the gas cloud and was extinguished.

'Sorry about that,' said a rather tight-lipped voice, 'My chap must have just missed the centre again. Can't understand it, my man is generally very accurate with these things. We'll just go around once more and get the blighter next time.'

Unbelievably, the Corvette made about four attempts to ignite the flare, with the Captain's comments becoming more and more terse. At one point we suggested that as they were busy guarding our sovereignty, he might like to leave it to the standby boat. But he wouldn't hear of it; obviously the pride of the Royal Navy was at stake.

As we now felt that things were getting severely out of hand, we decided to risk a call to the Captain suggesting that it wasn't the fault of his crew, but they might have more success if they fired the pistol at the extremity of the gas cloud. This would provide a sufficient mix of oxygen and gas to enable ignition to take place. He listened and to his credit, agreed to try once more. Now the whole platform held its breath as the vessel slowed; the sailor took aim and with a 'whoosh,' the gas ignited with a mighty roar. There was an enormous cheer from the hundred or so people lining the platform, the pride of the Royal Navy was once again restored and we wouldn't have to write to the Queen. We also felt that a potential Court Martial for the unfortunate sailor had been averted.

We thanked the Captain for his efforts and promised we would train the standby boat to light the flare in future, as we were sure the Royal Navy had more important things to do.

14

What's a McKissick?

It was a dark and foggy night (you can't fault that for an original opening) and the platform was waiting for a supply boat to deliver a cargo of construction materials. The boat was moving steadily towards the platform when you might say it rather misjudged its approach. As a result of this minor error, it ploughed through the steel jacket between the inside legs on the longest side and ripped out four major structural tubes, or 'members' as we in the trade call them.

It then demolished two more horizontal tubes just above the water-line and severely damaged one of the internal struts.

In order to grasp the scale of the problem, you have to visualise the enormous size of the steelwork used to build the jacket. The tubes, each weighing several tonnes, ranged from twenty to twenty-four inches diameter and were some twenty to thirty feet long. In addition, the boat also forcibly removed a thirty-six-inch diameter pipe with an underwater connection. In fact t collided with such force that it was more than half way inside the jacket before finally coming to rest.

Realising there had been what might be termed 'a navigation error' the boat then reversed its engines and, with a good deal of scraping and shuddering, removed itself from the inside of our platform. Those on board realised something was amiss when the platform rocked violently, crockery was thrown off tables, a snooker match was ruined and people fell out of bed.

As it was foggy, little could be done until morning when any personnel not directly connected with Operations were evacuated by a fleet of helicopters. Our Chief Engineer was flown out to take control of the situation and in the meantime, a specialist rigging company was mobilised. Their first job on arrival was to make their way onto the lattice structure and make safe the damaged tubes, a number of which were hanging loose and waving gently in the breeze.

This was a somewhat hairy operation as the only walkway within the jacket structure was a bit too close to the water—as I may have mentioned already. Scaffolding had to be rigged and temporary access ladders installed. In the meantime photographs were taken and sent to the Jacket Design Team so that they could begin to assess stability implications posed by the missing bits.

During one of our regular briefings, the Chief Engineer appointed me as the 'official liaison' between himself and the Rigging Manager. My instructions were simple: 'Find out what they need, and do whatever's necessary to get it as soon as possible.' (The job is technically known as a 'gofer'.)

Flushed with an altogether misplaced sense of importance, I climbed down to where the riggers were engaged in capturing the massive tubes, two of which were hanging down into the water at an angle of about 30 degrees. Gingerly approaching the Rigging Foreman, I introduced myself as the Company Representative charged with giving him and his team every possible assistance.

'Okay,' he said, 'I need a few items pronto, I can give you a list.'

Bracing myself against one of the undamaged tubes, I took out my notebook and replied, 'Fire away.' (Sadly, this attempt to sound competent fooled no one, least of all the Foreman.)

He then began to reel off a list of the items they needed urgently—tools, wire strops, clamps and so on. I began to relax, thinking that I could easily sort this out, either from our supply boat or, have them flown out by a charter helicopter.

However, just as I thought we were done, he dropped his bombshell: 'I also need ten McKissick snatch blocks, four with swivel hooks at one end and the others to be 'D' links. Oh I forgot, no less than two-tonne capacity.'

Both my newfound confidence and ballpoint pen disappeared in an instant. All I could think was, 'What the hell is he talking about?' Sadly, by this time, he had turned away to discuss some problem with one of his men and even given the opportunity, there was no way I wanted to show my ignorance by asking for clarification.

I couldn't write the last request down, as I had no idea how it was spelled. All I could do was make my way back up to the office whilst repeating *McKissick, McKissick,* like a mantra. Now, here on the page the word is spelled correctly, but you have to remember that when I received this verbal request, it was accompanied by the assorted sounds of a Force 5 gale, banging, hammering and shouted instructions in several languages.

So, on I went muttering, like someone with the D.T.'s, until I finally reached the offices. My next dilemma was who to ask? I was somehow sure everyone knew what McKissick snatch blocks were, except me. Pathetic really, but as I couldn't order the damned things until I knew what they were, there seemed to be nothing for it but to ask.

In terms of manliness and self-esteem, this was akin to having to stop your car and ask the way when you were lost.

The first person I tackled thought they had something to do with mooring boats and the next guy thought they were something you took for seasickness. I said 'McKissick is all one word, it's not bloody Mack ye sick.' He shrugged and walked away.

As time went on, I began to cheer up; I obviously wasn't the only one who didn't know what they were, but on the other hand it didn't solve anything, as I was still blundering about in the dark.

There was only one course of action left. I went to see Sid.

'Sid, what's a McKissick?'

'Ah yes, it's a block that's used as part of the Draw Gear on a drilling rig.'

I stared at him. 'Are you having me on?'

'No I'm not. McKissick's the name of a famous American manufacturer and I guess they also make the small pulley blocks you're looking for.'

Now it all made sense, 'Can you spell it please, Sid?'

'Certainly,' he said in a rather superior manner, and I was just about to make an ill-advised reply, when I realised a little show of humility on my part was expected.

The repair operation turned out to be very interesting. We had a huge barge (of film swapping fame) moored alongside, so that the riggers, welders, painters and assorted hangers on could play with great lengths of tubular steel.

Also in attendance was a diving vessel with a Remotely Operated Underwater Vehicle, known as an ROV, mounted at the stern. They were employed to carry out a series of unmanned surveys of the seabed around the platform to see where the broken members had landed and whether they had damaged our oil export line.

I was invited on board one day to witness the retrieval of the ROV and to see the film taken during the trip. This was most impressive, but paled into insignificance when I was invited to dine with the Engineers. Did I mention that an Italian Company owned the vessel? What civilised people they are; lunch was beautifully cooked and presented. White tablecloths and napkins dressed the table, proper cutlery and condiments abounded, but most incredible of all was that wine was taken with the meal.

Nobody on board thought this to be anything other than normal and conversation flowed freely with the wine during each course. As I was to find out later in my career, working with Italians is a lesson in politeness, civilised behaviour and kindness. This though, was my first exposure to 'style' and I loved it.

I mentioned the contrast to Sid on my return and he said, 'Bloody hell, can you imagine what would happen if we had wine with our meals?' Sadly, I knew what he meant; there is something in the British psyche which precludes moderation where drink (especially free drink) is concerned. This sad fact didn't stop us from asking from time to time, but our management were adamant, there would be no intoxicating drink on their platforms.

I asked Sid if I could transfer to the diving vessel for the duration, but he wasn't having it; I think jealousy was creeping in as, being the OIM, he wasn't supposed to leave the platform under any circumstances.

Throughout the course of extensive repairs to the jacket structure, the platform was subject to minimum manning restrictions. This meant that only Company staff remained on board to keep the plant running and continue drilling wells.

Until one day that is, we had to abandon the platform.

Thinking always of our safety, the Design Engineers had once again done some sums and had come to the reassuring conclusion that in our damaged condition, we were okay as long as the wave height was less than 65 feet and wind speeds were less than 50 knots. If such interesting conditions were likely, the OIM was required to evacuate all personnel. Doing so proved to be a good move on Sid's part as, in the event, wind speeds approaching 90 knots were recorded on the platform during our absence.

Preparations for us to abandon were initiated, production was shut down, the chip fryer was allowed to solidify and drilling had to pull the drill string out of the well, which took about seven hours.

Four helicopters were sent out to take everyone off, while all the time the wind speed was increasing. As it happened, the fourth helicopter couldn't make it back to Aberdeen because of hurricane force winds gusting to eighty-five knots. As a result, the passengers were diverted to a nearby drilling rig where they spent the night. Fortunately, the weather improved sufficiently the next day for the crew to return to Aberdeen.

All the lights were left on when we abandoned and, as it was essential to keep the navigation aids working, we left a generator running so that the batteries remained charged.

I was at home having free beer bought for me on the strength of my unassuming role in the drama and about three days into my off duty, when the phone rang. It was my Boss asking if I could come back up to the office as soon as possible. The next morning and feeling somewhat mystified, I arrived wondering what was in store—promotion, a pay rise, or the sack?

'I want you to go out to the platform, in advance of us manning up again. Take a couple of guys with you, carry out a survey of the situation and let me know whether I can mobilise the troops,' he explained when I arrived. 'Oh, and just to make everything legal, you'll be sworn in as a temporary OIM.'

Very reassuring.

Luckily for me, the 'soon to be' volunteers lived locally; Liam was the furthest away in Turrif whilst Bert and George lived on the outskirts of Aberdeen. I had a taxi at my disposal and zoomed around gathering up my waifs and strays.

Now, you may think that the lads were pleased to see me in the middle of their week off but this would be a dangerous assumption. Much to my surprise, they needed some persuading so, appealing to their fine sense of fair play and reminding them that 'You can't do enough for a good Company,' I awaited their affirmative replies.

They weren't long in coming. To a man, they told me what I could do with the Company, the Platform, fair play and my dubious parentage. I had obviously missed a trick somewhere and then the answer occurred to me: I had left out the loyalty factor.

Quickly regaining the initiative, I promised that overtime payments would start from early morning, not, as it was then, at two o'clock in the afternoon. Enthusiasm abounded.

It is humbling to find that once you have appealed to a person's better nature, they respond magnificently.

In addition to the gang, I obviously had to take Ken our Radio Operator with us and, possibly the most important person of all, a Cook called Les. The weather had abated but the sea still had that long grey swell that seems to linger after a severe storm has passed. We landed without any problem and I asked the pilot to sit on the deck until we were sure there was power to the deep fat fryer.

It was a strange feeling as we made our way into the accommodation, there were signs everywhere that people had left in a hurry; doors were open, clothes were strewn about and it was utterly silent.

Ken cheered us all up by saying, 'God, it's just like the Marie Celeste.'

We checked all the life support systems and found that the generator had stopped but the batteries, although low, still powered the emergency lighting. The other very noticeable thing was that the accommodation resembled a Chicago meat packers' cold store, as the boiler was also off.

A quick examination revealed that no new damage to the structure, so the chaps set to, ran up the generator and after a short time we were able to start a turbine, which provided enough power to make life bearable.

We reported to the Office that all was well, thanked the pilot and off he went back to Aberdeen.

In the meantime, Les was banging about with pans in the galley and brews of tea were appearing at regular intervals. With our major checks completed, food seemed to be urgently required, so we asked Les what could he produce and he suggested egg and chips. Now normally, a menu this basic would have been met with a tirade of abuse, but to a man, we thought his suggestion was inspired.

So, a short time later, we sat down to egg, chips, buttered bread, several bottles of sauce and pots of tea. Sometimes food just has to be of the right type, in the right place at the right time.

And, as a fitting reward for our efforts, we sat down that evening to watch an uninterrupted film and one we hadn't seen before—well not recently anyway. Full of chips and waxing lyrical, I thought this must be what being a 'North Sea Tiger' was all about; one thing was for certain, it was a hell of a lot better than being down a coal mine.

The following day we carried on checking and found only other some minor damage to the platform's lower section, consisting mainly of bent gratings and handrails. We were ready for the rest of the crew to arrive.

This they duly did the following day on a fleet of helicopters.

As Liam said later, he rather enjoyed having the platform to himself and had begun to resent other people tampering with his turbines. Bert, being singularly unimpressed, replied, 'The trouble with you is you're bloody power mad.' Not very flattering we felt, but reasonably accurate.

After waiting three more days for the weather to clear, repairs to the damaged jacket re-commenced, including the rather hairy job of filling a hollow cross member with cement—not the easiest thing to mix in the middle of the North Sea.

At last, after being shut down for some three weeks we were back in business and it was time to return to the excitement of mooring tankers to the Buoy.

15

Transports of Delight

Travelling to and from the platform was relatively straightforward; we went to the heliport, checked in, collected our survival suits, had a safety briefing and waited to board our flights.

A word here about survival suits: these are one-piece protective garments designed to provide the wearer with protection in cold water, where everybody who knows these things says you risk losing excessive quantities of body heat. There are many types of helicopter survival suit, made in a variety of rubbery materials; they are designed to cover the wearer from head to foot. Water is theoretically excluded by pulling on a long industrial strength zip fastener to close the garment, always assuming you had attended the necessary bodybuilding courses.

We collected our suits from the contractor who was tasked with their maintenance and distribution. Sizing was fairly arbitrary as they were designed (if that's the word we're looking for) to be worn over our normal clothes. Tall people, if they were unlucky in the availability stakes, were contorted into a shape reminiscent of the Hunchback of Notre Dame on a bad day, whilst small people disappeared altogether.

The suits were donned before boarding the helicopter and only removed when safely onboard or back on dry land. I use the word 'donned' in its loosest possible sense as they were fiendishly difficult to put on and even worse to take off.

Picture if you will the average guy in his jeans, check shirt, sweater and Clarks 'fit for comfort' shoes.

Over all this lot he had to pull on the survival suit, which, as it was made of a rubber-like material, didn't slip very easily over the existing apparel. This resulted in a good deal of pulling, sweating and swearing just to get the damned thing to waist height. Then came the next challenge:

how to pull the rest of it over your shoulders.

This manoeuvre involved some amazing dislocations of the spine and shoulders as our man heaved the suit across his back while at the same time trying to get both arms into the sleeves. Then, with one almighty thrust of the upper torso it was on and the sleeves were full of arms.

So far so good you thought; however the greatest challenge was yet to come, the unfortunate wearer then had to push his hands through an elasticated rubber wrist band. This was purposely made tight so that water couldn't get through and fill up the suit. The difficulty was that having managed to get the thing on this far the victim was exhausted and no amount of pushing would allow the wrist band to expand enough to get his hands free.

Finally, having made one last supreme effort, his hands were through the cuffs; he then found they were so tight that within a short space of time his hands began to tingle alarmingly. This was due to the fact that not only would the wrist bands keep water out, but also the necessary flow of blood around one's extremities would be similarly restricted. This led one to a choice of drowning in a water logged suit or risking the onset of gangrene in each hand.

As the average flight took about ninety minutes the symptoms were noticeable in that your fingers turned a nice shade of blue and you became unable to hold on to your newspaper.

I think the situation is called a 'conflict of design versus reality.'

Finally, once his victims had got the bloody thing on and collapsed exhausted, the sadist who was giving the safety briefing would make everyone stand and issue an order to 'Pull the zip up to its fullest extent and fit the hood.'

It was at this point you realised that you didn't care whether you drowned or not.

One problem, which was hardly ever discussed was that, except in flat calm conditions, there is much more risk of dying from drowning than hypothermia. It is said that in the North Sea in winter, there is a limited chance of anyone dressed in a standard helicopter survival suit lasting for more than about thirty minutes in the sea—even assuming he can get the zip pulled up and his hood fitting snugly, before entering the water.

Survival suits all depend on trapped air in the layers of the suit to provide thermal insulation. This in turn makes the suit highly buoyant. If however, the suit is too buoyant, then it will be impossible to make an escape from a downed, inverted, flooded helicopter.

As you can see in the serious bit above, they are an essential travelling accessory, but are not perfect. In the early days we had a sort of *laissez faire* attitude to suits, probably due to a misplaced feeling that 'big oil and gas men' didn't need to be mollycoddled.

In the early models, you had to take off your shoes and put your foot in the bottom of the suit which had a large flat pad underneath so that walking was reminiscent of a penguin with painful in-growing toenails. Another lesson quickly learned was that wearing either a kilt or a Burtons' three-piece suit was not the most appropriate garment for stuffing into a survival suit, although it was fun watching someone try.

When we finally boarded the helicopter, we used to do something, which, on sober reflection, was lunatic in the extreme. On finding a seat we would fasten a packed life-jacket around our waist and then with a supreme effort and occasionally some help would remove the top of the suit until it sat in a mangled heap at waist height. This had a number of fairly obvious disadvantages should the helicopter inadvertently land on water. Putting the suit on, pulling on the hood and zipping it up fully would take about ten minutes; only then would the life-jacket be revealed, ready to undo, pull over the head and prepare to inflate.

As you can imagine it only took one near miss for the painful message to be brought home; image must sometimes be sacrificed in order to stay alive.

The final delight was the result of a brilliant idea by someone who had never been offshore—that it was only necessary to have one set of suits. The lunatic theory was that on landing, we would simply remove our suits and hand them over to the guys on the platform waiting to go home. This assumed that for every person arriving there was an equal and opposite clone waiting to be kitted out in your suit. Strangely enough this was not always the case and it was not unusual to find yourself, having just stepped onboard, being forcibly disrobed by someone of roughly the same size.

Oh, and it might be worth mentioning that wearing one of those suits for upwards of two hours caused the wearer to sweat *a lot.*

Back to our mode of transport. This was a Sikorsky S61 helicopter fitted with canvas seats and we thought it to be noisy and uncomfortable. Until that is, I went further north and we travelled variously in a Boeing Chinook, a Eurocopter Dauphin and a Super Puma. They were all so cramped compared with the Sikorsky that I often wished I had been less hasty in criticising its admittedly frugal attributes.

As I had never even been close to a helicopter before coming offshore, I was fascinated by their amazing flexibility, which meant that they could fly almost anywhere. However, it also means that flying the machine is complicated. The pilot has to think in three dimensions and must constantly use both arms and both legs to control the machine. Piloting a helicopter requires a great deal of training and skill, as well as continuous attention to the machine. I learned all this from Allan who I am sure, spoke with due modesty.

He also told me that there are three things a helicopter can do, which an aeroplane can't. It can fly backwards, the entire aircraft can rotate in the air and it can hover motionless above the ground. I thanked him for this reassuring information and assured him he was welcome to do all of those things, as long as I was not on board.

Sadly for us, there were times when we needed to get to and from the Buoy despite our trusty Bolkow being out of service for one reason or another. Notwithstanding this so-called minor setback, tankers still had to be attended to as 'them onshore' seemed to be overly interested in maximising oil revenue.

Normally, we flew over to the Buoy, but getting back was often another matter, or should I say, adventure. This good news was often communicated to us just as, at the end of a day in the bracing wind, we asked Allan to come and fetch us in the mean machine.

'Sorry chaps but the Bolkow isn't serviceable at present; you'll just have to make other arrangements.' The fact that these 'arrangements' were never defined meant some time was wasted in plotting all sorts of revenge on Allan, his mechanic and anyone else we could think of. However, if we wished to have dinner, watch a film, sleep in a bed or curse Allan in person, our options were generally limited.

The favourite way off, and I use the word 'favourite' in its most imaginative context, was to ask the supply boat to take us aboard and deliver us to the platform. Now Supply boats are a funny shape, in that all the working parts are stuffed up at the front, or bow as we sailor types call it. This leaves space for a long, flat open deck stretching for about three quarters of the length of the vessel. Whilst there are rails along the sides, there is a big open space at the stern, generally with a huge steel roller running from one side to the other. This was used to allow anchor chains or mooring ropes to slide easily into the sea. You may perhaps see that leaping onto the deck required us to miss the roller if we didn't want to find ourselves deposited nice and smoothly into the water.

Now, the leaping itself was hairy in the extreme and required us to have total faith in the seamanship of the Skipper. Fortunately, whilst I was there, it was always the same boat that carried out the preparatory work in running out the messenger line ready for mooring. This meant that we knew one another as we had, on many occasions, observed his skill in manoeuvring the vessel and had also built up a friendly relationship with the crew.

Taking us off the Buoy required the three of us to make our way along the steel structure and stand on the wooden fenders. This sounds easy, but generally it was late on a winter's afternoon, it was becoming colder, the Buoy was moving up and down and side to side in the waves and we were each balanced on the flat top of a fender, a bit like those seagulls you see standing on flagpoles, but with a lot less confidence.

The next part required complete faith in the Skipper as he came astern and slowly approached the fenders. Now, what he couldn't afford to do was hit them, as even a gentle nudge from a 500 tonne boat would throw us off balance and spoil our day. Still, we joined for the adventure.

The Skipper inched his boat towards us and turned on his powerful deck lights. These were mounted on the bridge and shone down onto the flat deck. This meant that he could see us, but as the boat moved upwards on a wave, we were dazzled.

I remember asking him one evening if he would turn off his lights, his reply was brief and to the point: he needed to see us and he didn't care a damn whether we could see him or not. Fair enough.

The final stage was governed entirely by the Skipper who, when he had inched towards us and was at his safe limit, shouted a very succinct instruction over his loud hailer.

The instruction was in two parts. Firstly he said, 'When I shout jump, I want you to leap onto my deck.'

The first time we heard this we stared at each other in something approaching horror. Bert wondered, 'Is he bloody serious?'

I said 'I think so—how else did you think he was going to take us off?'

'Sod this,' put in Liam, 'I am definitely going to kill Allan as soon as we get back.'

'You mean if we get back,' added Bert, continuing in his optimistic vein.

Just then, we became aware of the second part of the instruction.

The Skipper, watching the wave cycles and his stern, shouted 'jump', just as the rear of the boat was lifted by a wave to about within five or six feet below us.

As we geared up to jump, he let rip with a final command, 'Now!'—And we all stayed exactly where we were, just as the boat plunged downwards once more.

This mutinous behaviour didn't sit well with the Skipper as he moved the boat away from us and for a while we really thought a night on the Buoy was in prospect when, with a furious churning of water, the boat reversed towards us once more.

This time we gathered ourselves for the leap, realising it was our last chance of supper that night. Our eyes were glued to the stern as once more the boat dropped downwards and then quickly rose upwards.

As we mentally prepared ourselves for the big event, the Skipper's voice blasted out of the tannoy.

'Bloody jump—now!'

The outcome was hardly dignified; we landed in a terrible heap of arms, legs, overalls and helmets as we rolled around the deck. Of course, what we hadn't realised was that the instant we literally hit the deck, the Skipper would go ahead at maximum revs in order to get away from the Buoy. He was thinking, I suppose, of the cost of damage to his boat, the Buoy and the potential loss of oil. We, on the other hand, were only concerned with staying alive and not missing a needle snooker match scheduled for that evening—it's all about priorities I guess.

His deck hands, who were as sure-footed as monkeys (this is meant as a compliment,) helped us to our feet and the voice of the Skipper boomed out again, 'All right lads, off we go to the platform.'

Ten minutes later, we were under the lee of the platform and a basket, which looked like a giant round rubber dinghy with netting fixed to the sides and meeting at a point to which the lifting eye was fitted, was lowered down by our crane. The idea was that as the crane lowered the basket onto the deck of the boat; you stepped smartly onto the rim, grabbed hold of the netting and were whisked away before the deck of the boat heaved upwards.

Once again timing was everything, because if the rope slackened and the crane driver started to reel in, you could find yourself being pulled along the deck towards the open stern, with every prospect of being dumped into the sea. By then of course, the slack in the rope would have disappeared and you would be hurled into the air. Only exciting if you had some sort of death wish.

So, having learned a hard lesson, we timed our steps to perfection and were winched up onto the helideck. And people think that Blackpool Fairground is scary—hah!

It is interesting to consider that we told no one onshore about our various leaping acts. They in turn, never asked how we managed to moor tankers without the Bolkow. And we never thought to bring the subject up at safety meetings.

Thirty years later, I look back on this lunacy and the thought of what we did still makes me shudder. I suppose we did it because we didn't want to let the side down, or for something far less altruistic, like money.

Could it have been that we three enjoyed the adrenalin trip? Or were we simply too slow to appreciate the seriousness of the situation? Who knows? But it seemed like a lot of fun at the time. I do know there is nothing in the world that would make me leap onto the back of a supply boat these days. Is this due to sense, cowardice, old age or the lack of companions willing to join me? I don't intend to find out.

On another occasion, we were once again *sans* Bolkow but this time the sea was somewhat calmer. Our supply boat was engaged in retrieving the messenger line after the tanker had left (bearing oil for the Chancellor) and couldn't assist in our evacuation.

However, anchored nearby was a Diving Vessel carrying out some remedial work to the subsea pipeline. We called him on our radio and asked whether it would be possible for them to take us off the Buoy before nightfall.

'No problem,' said the officer on watch, 'We'll launch the Zodiac and come over to take you off.' The Zodiac is an inflatable dinghy with an outboard motor that was used to retrieve divers from the sea and, as their offer seemed better than staying put, we didn't enquire further, which as it turned out was another mistake.

Fortunately the vessel was anchored fairly close by and as we watched, the inflatable was lowered and two crew-members set off towards us. They are fast boats and bouncing across the swell it soon careered to a stop close to the fenders. This time, as the dinghy was low in the water, we had to make our way down a rusty iron ladder which was fixed (we hoped) to one of the fenders.

A seaman held onto a rung of the ladder whilst we climbed down and simply fell into the dinghy once again in an untidy heap (this seemed to be our lot in life.) Only then did we realise that the sea was in far from 'mill-pond' conditions. We were being bounced around in the bottom of the dinghy with all thoughts of impressing the crew with our seamanship long forgotten.

Meanwhile, the sailor holding the ladder was having something of a struggle to keep the craft from bending itself around the fender.

Fortunately, even as we lay in a tangled heap in the bottom of the dinghy, the other sailor opened the throttle and we shot off towards the diving vessel. This sudden acceleration thwarted all our attempts to sit up, admire the view and act like the old sea dogs we fondly imagined ourselves to be. Instead we lay against the inflatable sides with our limbs entwined in a most disgusting manner and prayed for salvation.

Little did we know.

On arrival at the ship, we began to realise just how big it was when you were in a silly little rubber boat staring up at a wall of rusting iron. We were again held to the side by the seaman as we surged up and down in the swell. Then we realised that the only way on board was to climb a rope ladder, which was hanging down the side. However, as both the vessel and our dinghy moved in the swell, the ladder swayed away from the side of the ship by about six feet, only to clatter back again as everything moved in the opposite direction.

The crewman on the engine then said the magic words, 'Right lads, off you go and don't hang about, the sea's getting up.' This, as it turned out, was another of those throwaway seafaring phrases designed to reinforce our already growing feeling of despair and hatred of the Company.

We then began to realise the expression 'hang about' although accurate was, as far as we were concerned, unfortunate.

If you are bored one day, try to stand up in a moving inflatable boat, pause, put one foot in the bottom rung of a rope ladder whilst at the same time grabbing the rung above your head, then lift yourself clear of the dinghy. Another strategic error was that we were also clutching our insulated overalls under one arm.

It slowly dawned on us that fundamental to a successful career in climbing rope ladders up the side of a ship is to have two free hands.

Thinking quickly, we stuffed the overalls down the front of our boiler suits, which, unfortunately, had the effect of forcing us into a sort of reversed Quasimodo posture.

Not only that, but never having climbed a rope ladder before, we failed to realise that there is an art in so doing.

When you pull on the rung above your head, the blasted ladder leaves the side of the vessel and you find yourself in a most alarming position where your feet are pushing you in and your hands are pulling you out. At this point you are arched backwards and staring up at the deck rail some miles above.

To compound this farcical situation, if you grab the upper rung with say, your right hand placed to one side, the ladder lurches rapidly to the left. This additional excitement means that you are now making no upward progress whatsoever but are in a strange yoga-like posture, for which the human body was never designed.

This is fine except for the fact that there are people below who don't understand what the hell you are doing and are yelling for you to get a move on. The real adrenalin boost comes when the next person makes a bid for the ladder and is now hanging on below you. He is just about to find out what you have learned and of course, he thinks that his predicament is due to you moving the ladder. This results in much useless yelling and accusations along with the realisation that the third member of the team is still in the dinghy, totally unaware of the excitement to come and is anxious to begin what he thinks is a simple climb up a ladder.

Worse still, you are conscious of the fact that you are being watched by two experienced seamen who just want to get on board and stow the Zodiac. Talk about embarrassment.

However, there is nothing like a combination of panic and looking a complete prat in front of an audience to get one moving. So slowly and with a number of Tarzan-like moves, we made it to the top of what seemed to be a vertical cliff face made of freezing cold, wet iron, and onto the deck of the ship.

You may have noticed a common denominator in the outcome of our outlandish modes of transportation, which is the fact that we generally fell onto things, into things and off things. Our arrival was generally far from dignified and it was somewhat humbling to find you were lying in a crumpled heap at the feet of an immaculately turned out First Officer, whilst at the same time trying to appear dignified and fit to be in control of a producing Platform.

It seemed to me that I had make far too many introductions from a horizontal position, still entangled with the other two, whilst at the same time making futile attempts to stop them from cursing everyone in sight.

Gratitude, I am sad to say, generally occurred to us too late to thank our rescuers, but I believe we at least gave them all a great deal of entertainment.

Getting to and from home in the early days was also something of an adventure. The normal way was to get the train from Aberdeen and, as I lived in Cheshire at the time, I could alight at Warrington station, which was only about eight miles from home. However, as I usually went into the office before setting off, I missed the afternoon departure and ended up on the last train, which left at about 11pm.

For some unaccountable reason, this train would stop at Carstairs (near the Mental Institution, which in the light of what then happened, seemed appropriate.) Here, the rear section would be separated from the front before making its way towards Cheshire. It was, therefore, very important to be seated in the rear section and to be wide awake, whilst the shunting activity was in progress. If, that is, you wanted to end up in the right part of the country in the early hours of the following morning. I never did find out what became of the front end of the train, but I think it may have been the inspiration for the film *Oh Mr Porter*, starring (if that's the right word) Will Hay.

The other slight snag was that the late train didn't stop at Warrington, so I had to get off at Wigan, which was a considerable distance from home. Not, under normal circumstances, too much of a problem until you realise it was four-thirty in the morning when I stepped onto a deserted platform. This meant that my long suffering wife had to get up in the middle of the night, make her way to Wigan station and look for a knackered husband standing under the exit sign, which as there was no one else there at the time, wasn't too difficult.

This worked reasonably well (for me, at least) until I confused things by occasionally travelling on a scheduled flight to Manchester Airport. As I never knew in advance which mode of transport I would be using, my better half had to wait for a last minute call from me in order to organise the pick-up arrangements. Remember this was in the dark ages before mobile phones were invented.

As she had to get up in the middle of the night to drive to Wigan, she was generally on 'auto pilot' when setting off. This wasn't too bad, as there was very little traffic at that time of night until, on one occasion, she found herself on the outskirts of Manchester Airport instead of Wigan station.

This small navigation error then required her to find a public telephone, find the number of Wigan Station and ask the night duty Stationmaster if he could see a lonely, tall, fed up looking chap standing outside. Good enough, he did find me and was able to advise me to stay there, as my wife was on her way across Lancashire to pick me up. Good advice, but then, where the hell could I have gone anyway?

Now, younger readers, brought up on *Thomas the Tank Engine* stories, will be surprised to learn that in the good old days of British Rail, there were real, kindly, helpful Stationmasters; their disappearance is due, I am told, to modernisation and 'improvements'.

On another occasion, my wife was on her way to the station when a police car stopped her. The policeman wanted to know what she was doing out at that time of night so, being of a suspicious nature and with nothing better to do, made her open the boot. Quite what for we never found out. My wife managed to convince him that she was not a courier for a major drugs baron and he, missing some key evidence, waved her on her way.

So you see, the thrills and adventures of working offshore were not just confined to me, except that my wife's participation seemed to involve driving randomly around Lancashire and Cheshire, mainly when everyone else was asleep.

The best though, as they say in musicals, was yet to come.

I was lamenting to Allan one day that travelling home was proving to be something of a pain, not only for me, but for my wife, who had to miss most of a nights sleep, get three children off to school and then lecture to eager students all day, at her college.

Allan said he had a contact in the flying world who might offer me an unofficial lift down to Liverpool Airport (now known, unaccountably, as John Lennon Airport.)

Apparently every evening, a Vickers Viscount aircraft belonging to a company called Alidair, flew from the old side of Dyce Airport down to Liverpool. Its job was to pick up newspapers and then make the return journey back to RAF Lossiemouth, which was even further north than Aberdeen.

There were just two crew members and Allan said I should to go to their office and ask whether there was any possibility of a lift to Liverpool. As it transpired, I was successful and, having signed a waiver absolving the company of any liability should the plane be inadvertently shot down, was told to follow the crew out to the plane.

Interestingly, the Viscount was the world's first gas-turbine 'turbo-prop' airliner. It was renowned for its comfort and low interior noise level, something to which I am able to testify. Allan, (again) told me that the first one entered service in 1953 and the last in 1964. Initially it carried fifty-eight passengers but later this was increased to seventy-six (obviously with the offshore industry in mind.) The engines used were the Rolls Royce Darts giving a cruising speed of 320mph with a range of 1400 miles. I told you it was interesting.

I duly boarded the plane and as I was the only passenger, picked my seat and made myself comfortable. At about five in the afternoon and without much in the way of a safety preamble from the crew, we were on our way to Liverpool. Once again, I had to let my wife know what was happening and when, so I had to be able to make that last minute phone call. Happily, the office staff always obliged.

This new travel arrangement worked extremely well for a number of months and there were never more than two or three chaps on board who, like me, had somehow discovered the unofficial transport system. I remember on one occasion, I was the only passenger and the co-pilot asked whether I would like to sit in the cockpit and watch them fly the plane. As I had read my book and there was no in-flight magazine, I spent a fascinating journey watching levers being pulled and knobs being twiddled. An added bonus was that the flight took place in daylight and being on the flight deck, as we seasoned flyers say, you could pick out notable landmarks like Blackpool Tower and less notable landmarks like Manchester.

By now, any lingering nostalgia for coalmines had faded into obscurity; this was the life. Did I mention the other significant advantage of this arrangement? It was free.

For some months our arrival time was always perfect, allowing my wife to meet me early enough to go for a Chinese meal. Until one day, both the timing and destination went slightly wrong.

On this particular occasion there were three of us on board, one guy off a drilling rig who had flown with us before and a friend he had brought along to share in our mutual good fortune. As I had by now become somewhat blasé about flying in this way, I was able to sit back and 'earwig' on their conversation.

They were both from Liverpool and they spoke like Jimmy Tarbuck on speed. As the newcomer was very excited at the prospect of an executive flight home, he smiled a lot and kept thanking his mate for the opportunity. This outward sign of childlike gratitude made me feel even more superior, as only a seasoned traveller on a 'freebie' can.

Soon we were airborne and our friend's boyish enthusiasm continued, interspersed with repeated, and as it turned out premature thanks to his benefactor for sharing this incredible good fortune with him.

It was March and it rapidly became too dark to see out of the windows, so slowly our friend relaxed and we all nodded off.

We awoke some time later to see the co-pilot clutching a torch and peering first of all out of the starboard window, then moving across the cabin to peer out of the port side. Somewhat puzzled, we watched and waited but not a word was said and he went back onto the flight deck. We had just settled back again, when he reappeared and went through the same torch peering routine.

The third time it happened, I thought 'sod this' and asked him why he was nipping in and out shining a torch out of the windows.

He then came out with one of those classic understatements that you really don't want to hear.

'Not to worry but we think there may be a problem with the undercarriage. We're not getting a signal to confirm that it's engaged ready for landing and I can't see anything through the windows.'

And with these words of comfort, he switched off his torch and disappeared back on to the flight deck.

There was now what may be termed a pregnant silence and then our excited friend turned to his mate and said those immortal words:

'There's another fine mess you've got me into, Stanley'.

You really had to be there to appreciate just how well timed and appropriate this classic one-liner sounded, and I dare say was it never better received, even by Oliver Hardy. However our laughter was of the more nervous variety.

The co-pilot then reappeared and staring at three idiots rolling about in our seats said, 'Sorry chaps, but as we can't confirm the wheels are down, we can't land at Liverpool because they don't have the necessary emergency support systems for a potential belly landing. So we're diverting to the East Midlands Airport as they are geared up for this type of thing.'

With that he was gone again.

We now sat silent and upright in our seats as we felt the aircraft turn away from Liverpool and head out to the East Midlands. What on earth had we let ourselves in for? Where was my seat belt? And what the hell did he mean by 'This type of thing?'

It then dawned on me that about now, my wife would be arriving at Liverpool expecting to pick me up as usual. The problem was, as this was an unofficial flight, it officially had no passengers, there was no flight list and obviously no information regarding a change of course would appear on the bulletin board in the Airport lounge.

One thing at a time I thought, let's see if we land in one piece before worrying about contacting anyone.

On approaching our new airport, I was reassured to observe that the pilot had obviously seen *The Dambusters*. He banked slowly round just as the hero had done when approaching the Mohne Dam. It was also clear to me that the co-pilot must have been calmly issuing the 'Steady skipper, keep her on this heading, slowly does it.....now' instructions, as the aircraft landed nice and gently on the runway.

Fortunately, the wheels were down, and red engines and flashing yellow lights surrounded us. We all breathed a collective sigh of relief. Sadly, I thought later, this had been yet another missed opportunity to be an unassuming hero.

Back to reality, we now stood outside a strange airport wondering how to get home. In the meantime my wife, having realised the plane was long overdue, managed to make contact with the Alidair representative, who gave her a cup of tea and the bad news.

All she could now do was wait until word came back that we had landed safely and then make her way home. Worst of all was the realisation that there would be no Chinese meal.

After standing around the terminal looking lost and forlorn, we managed to travel from the East Midlands to Liverpool by cadging a lift from the co-pilot, who had to return and organise the delayed newspaper deliveries to the far north. I then had to ask my wife to come out again and fetch me. Only the brave and so forth.

Looking back, I can see we had a number of escapades where a mobile phone would have been very useful, but in those dark days long ago, we had no idea that such wonders were possible. At least we didn't have the vexed problem of having to use our thumbs to write letters.

16

Just a Slave to Fashion

You have to remember this was 1975 and for most men, the word 'fashion' was something that only concerned women. 'Fashion' was something you saw in glossy magazines and most men firmly believed that the contents of their wardrobe would be 'in fashion' every ten years or so.

However, it would be wrong to assume that we were just a 'bunch of scruffs' whilst going about our duties. Simply because we looked pretty awful most of the time didn't mean that we weren't as aware of Mary Quant as the next man. It is true that some of the chaps didn't know what pop group she was in, but that didn't mean we lacked taste and discernment when it came to dress code.

Many of us were already completely *au fait* with the cutting edge of male fashion; after all this was the era of Kevin Keegan's perm and Henry Cooper advocating the liberal use of 'Brut'. Cardigans were the casual garments of choice, generally brown, or for those with a colour co-ordination problem, a rather more daring beige with stripes.

It was also rumoured that some offshore workers secretly preferred Sacha Distel to the Rolling Stones, but this was never proved.

Aspiring to be a 'Big Oil and Gas Man' meant that certain items of apparel were, as we say in the business, *de rigeur.*

For example, a lumberjack shirt, jeans and trainers immediately identified the wearer as one who had his finger on the fashion pulse. The shirt would be open at the neck revealing, on those of a simian bent, a quantity of chest hair. Some of the newer recruits made the fatal error of undoing just one too many buttons on the shirt; one should never flaunt unsightly flesh.

You could always distinguish a visitor from a regular offshore man, in many cases the visitor would turn up to the heliport in a collar and tie and

I recall one chap who even arrived wearing a Harris Tweed jacket.

Sartorial gaffes such as this were looked upon with pity by the hardened members of staff and some of the less feeling among us found it to be a great source of amusement. This was enhanced by the knowledge that pretty soon it would all have to be squeezed into a survival suit. What fun.

Worse, they generally arrived with a suitcase, instead of the usual nylon holdall. The case would typically contain brand new overalls, gloves, safety helmet, protective goggles and safety boots, none of which had ever been worn. We always enjoyed the look of horror on their faces as, having carefully put the case down by the helicopter's hold, they witnessed it being flung into the back, followed by a stack of drilling equipment landing on top.

That this sensitive method of baggage management prevails today in all major airlines says much for the united front prevailing amongst handlers. I guess there must be an international code of practice on baggage destruction somewhere.

Having survived their first trip offshore, these tyros were now faced with the problem of re-filling the suitcase ready to go home. Their now filthy, oil-stained boots and overalls had to be laid either under or on top of six or seven unworn white shirts and accompanying ties, which had been packed in the mistaken belief that we changed for dinner.

You see, there were many very important rules to learn regarding the proper way to travel offshore. Stumbling up and down exposed iron stairs from the helideck, in a force six gale, whilst clutching a now somewhat out of shape suitcase, briefcase and Macintosh, did nothing to help the newcomer to blend in.

A gold chain worn around the neck, never on the wrist, would complete the 'Offshore Tiger's' ensemble. It was most important to observe the 'gold chain protocol' when dressing. Sovereigns were never worn; they sent a signal that the wearer was confused about money. Similarly, crosses and items of a religious nature were frowned upon, as they indicated the wearer lacked confidence in his ability to stay alive without divine intervention.

The same could also be said about the St Christopher medallion, although a number of misguided wearers, on being challenged, muttered that their wife or girlfriend gave it to them 'for luck'. When they were told that St Christopher didn't actually exist the desired 'rough, tough' image was further dented and by the following trip St Chris had generally been consigned to religious limbo.

This was yet another example of the need to carry out detailed research since a failure to do so could invoke derision and ruin the desired effect.

Earrings were also considered somewhat suspect with regard to signals the wearer was sending. This was borne out by a comment made by an American comedienne who believed that men who have a pierced ear are better prepared for marriage—they have experienced pain and bought jewellery.

There was a contention that, in some cases, earrings were worn only when working offshore, as there was no way the wearer, having moved back into domestic mode, could go into his local pub with one (or two) *in situ* and live to tell the tale. Remember, David Beckham hadn't yet been invented.

Jackets could be one of two types. The first was designed to be worn in inclement weather, it was long and insulated with a built in hood, tastefully finished with a piece of mock fur around the edges. It was fastened by an industrial sized zip reaching from the bottom of the jacket to the underside of the chin. Obviously the fashion conscious among us never closed it all the way to the top—whatever the weather.

The other outer garment was the ubiquitous fleece jacket. The early type with which we were issued was navy blue with a zip running throughout its length. The outer material resembled a sort of stiff sacking, whilst the inside was made of a polyester pile containing enough static electricity to light a street lamp. It had two 'patch' pockets and elasticated cuffs.

Longevity was not high on its list of attributes and after a couple of wearings, the cuffs would have expanded to about twice their original size, the zip would be slowly parting company with the stitching and the outside sacking would be wearing thin at the elbows and shoulders.

A tragic error would be to have the jacket washed whilst offshore. This would result in a radically altered garment that would perfectly fit your youngest child.

In our ignorance, we were relatively content with our apparel, until one day a Diving Superintendent came on board wearing a fleece jacket, the like of which we could only dream of. It was again navy blue with a full-length zip, but that's where any similarity to our jackets ended.

His jacket was long; the cuffs fitted snugly, the pile was on the outside as well as on the inside, but most of all, discreetly embroidered on the chest, were the magic letters 'HH'. They were in two-tone blue with the edges of the letters just overlapping one another.

We stared in disbelief; he was actually wearing a 'Helly Hansen', a garment fashioned with love and attention to detail, by the famous Norwegian Company. This was an item of clothing on a par with suits hand-made by Hardy Amies (whoever he is).

On seeing the jacket for the first time, the more sensitive souls among us had mixed feelings; a touch of jealousy mingled with pride, to be in such close contact with this ultimate in fashion statements.

Throughout all this adulation, the diving man had a look of faint bemusement, mingled with what I took to be pity. It's awful to be a second-class citizen in matters of such importance. But we soon regained our composure, agreeing that having to dive in the North Sea was too high a price to pay for even the ultimate in fleece jackets. You see how brave and resilient we were.

Nevertheless, one person did commit the ultimate in bad taste by offering to buy the jacket from the Superintendent. We guessed he was probably carried away at the thought of the 'kudos' he would have entering his local, while casually wearing such a priceless garment. What a prat.

Boots were the next sartorial item on the list; ours were of the slip-on, calf-length type made of off-cut brown leather, with a flimsy loop at the back as an aid to pulling on the foot. The loop could only be used once, as the next time it invariably snapped.

The boots had steel safety toecaps which were covered in the thinnest film of leather imaginable. This meant that very soon the leather frayed and exposed the steel cap underneath. Interestingly, no one thought a piece of rusting steel poking through flapping bits of leather was in any way unfashionable.

Personalisation of boots was generally restricted to the introduction of the wearer's initials in ballpoint pen on the outer top edges, some in a sort of pseudo-gothic script with no two letters of the same size.

And then one day, we had a visit from one of our Construction Managers, who, prior to entering Sid's office, observed the unwritten protocol and took off his boots and left them at the door. Pretty soon they became the subject of much pointing and whispering; word went around and shortly the corridor was full of people trying to get a good look at the boots.

Then one brave chap risked all and picked one up for closer inspection. The rumour was confirmed; there, embossed on the side for all to see, was the winged logo. He was actually holding a genuine 'Redwing' boot.

Boots such as these were from fashion heaven. First they were as worn by big-name American oil and gas men like the famous Red Adair. Second, we Brits could only dream of owning this ultimate example of offshore footwear; it seemed to be that our lot in life was one of envy, not a pretty sight.

Who says the British male has no dress sense? Just about everyone I suppose.

It was obvious that our Company was unaware of the negative signals being sent around the oil patch as a result of their failure to provide proper sartorial support. This of course meant only one thing; the subject would have to be raised as an urgent item at the next committee meeting.

I won't bore you with the protracted discussion this provoked, but I realised early on that Sid, who wore slippers and a cardigan whenever possible, was hardly the going to pursue our request for Helly Hansen jackets and Redwing boots with any degree of enthusiasm. His rather spurious argument was that our Company would not be interested in 'following the herd'. As we had no idea what the hell he was talking about, we realised there was little chance of success and reconciled ourselves to second-class citizenship in the sartorial stakes.

Dickies overalls were worn complete with as many work stains as possible; new overalls were given several washes, both to remove the stiffness and to provide just the right degree of 'fade'. The ultimate in wearer 'chic' for overalls would be to have the ruler pocket torn. This was not difficult to achieve as the pocket was located at the side of the leg and was prone to catching on valve handles, snooker cues and the edges of desks.

In a similar manner, gloves had to have a well-used look and be kept hanging out of the back pocket of your overalls. If you were really lucky, you could cadge a pair from the Construction lads that had genuine leather wrist guards. Obviously the main criteria for displaying optimum glove status, was never to be seen wearing them.

Safety helmets were plastic and colour coded to denote the Company for which you worked. To provide wearer credibility they had to be scratched wherever possible, with ear defenders gripped around the hat, just above the rim. Some misguided people stuck various contractor badges on their helmet in a pathetic attempt to indicate the wearer's experience in international oil and gas projects. This was a cardinal error since when questioned, the berk would soon show he had no idea what the various companies did for a living.

The ultimate goal for all discerning safety helmet aficionados was to obtain a metal one. Our Company didn't issue this type, but if you played your cards correctly and made friends with a Construction Supervisor, you could often obtain one. Sometimes they would even cut your name into a copper tag and epoxy it to the front of your helmet; this meant of course, that there was an opportunity to polish the copper, thereby adding a further uniquely fashionable dimension to your ensemble.

The helmet of helmets, the ultimate choice in fashionable headgear for any budding oilfield fashion victim was also made out of aluminium alloy and had a real leather chinstrap; but far more importantly, it was an MSA helmet.

So what, you may ask. Well it is important to understand that the letters MSA are the initials of the Mine Safety Appliance Co, and these helmets were made in Pittsburgh, Pennsylvania. Just ask yourself, with a manufacturing address like that, how much more credibility do you need?

The most memorable example of helmet decoration I ever saw was one belonging to a very pleasant Toolpusher named Gus who joined our platform after the resident maniac was removed to a secure unit for deranged Toolpushers. Gus had recently completed several years' duty in the Far East and having been well liked by his Indonesian crew, was given a decorated MSA helmet as a going away present.

One evening he opened his cupboard and showed me the finished product, which was genuinely astounding.

His crew had given the helmet to a craftsman metalworker who earned a living making fancy metal plates and dishes for tourists. The result was a polished helmet intricately decorated in embossed flowers and myriad patterns, the like of which I had never seen before. At the front of his hat, the artist had picked out Gus's name in beautiful gothic script. Obviously, it couldn't be worn in service, but as a memento and a tribute to the metal-beater's skill, it was superb.

Another 'must-have' accessory was the Briefcase, and not just one of the ordinary flat affairs as favoured by debt collectors and accountants, but an identical model to those used by the Helicopter Flight Crew. This was rectangular in section with a fold over lid at the top, through which the handle was located. It was big, heavy and in the main, useless.

Fashion junkies who spent a great deal of money purchasing such an object were immediately at a disadvantage. Unlike our soft holdalls, which could be flung unceremoniously into the hold, the Flight Crew briefcase had to be carried in the cabin where leg room was already at a premium. This meant that the case had either to be balanced on the knees for the duration of the flight, or it had to be placed at the side of the seat. As the latter method blocked the aisle and the case on the knees stopped the owner reaching his life-jacket, either location was somewhat fraught and a sense of regret at the hasty purchase could often be observed.

The other small problem was that the new owners had no idea what to put in the damned thing. Real Pilots of course, carried their flight plans and instruction manuals within; our heroes struggled to fill the interior with spare socks (never worn), the Sun comic, assorted Costa Brava holiday brochures, a Wilbur Smith paperback and a bag of mintoes.

Another fashion no-no (as with the safety helmet,) was to cover the sides of the case with various company stickers. These could again be obtained from Drilling Contractors and tended to be large and colourful with Company logos and mission statements around the edges. As a macho activity, it was akin to collecting train numbers or Butlins Holiday camp badges.

I remember Liam standing near to one such 'sticker collector' at the check-in desk and staring rather fixedly at his highly decorated case. After quivering for a minute or two, he asked him what all the badges were for.

The guy looked a bit sheepish and then defensive and muttered that he had put them on simply to enable him to identify his case, as it came around on the luggage carousel.

I could see that this explanation was not sitting too well with Liam, who replied, 'Wouldn't you be better tying a bloody big red ribbon around it instead? Then you could spot it a mile away.'

I intervened by trying to change the subject, a move that satisfied no one, especially Liam who kept muttering 'What a bloody prat,' under his breath.

Finally, all the offshore power dresser needed to ensure sartorial status was a wristwatch. Now there was one watch coveted by all and afforded by none and it was of course, the Rolex. It is said that one can only aspire to greatness; it was certainly true when applied to actual ownership of this particular timepiece.

The nearest most people got to the Rolex was a claim that they had once met a guy in a pub who had been in the Far East and was actually wearing one. For us, the affordable watch of choice was probably the Seiko; they had a certain 'street cred' and were made in a tasteful stainless steel. What's more they kept good time and had saved many an owner from missing the first crucial ten minutes of a film.

Some people made the fatal error of going over the top and confusing 'add-ons' with good taste. Watches could be obtained with numerous dials and buttons to tell depth of water, phases of the moon, the time in Ulan Bator, conversion from Fahrenheit to Centigrade and the average temperature in Bangladesh. All very useful stuff, but the watch became so bulky it resembled a cash register strapped to the wrist, especially if one of the buttons caused the face to light up. The sad reality was we all knew the wearer was trying to convince the gullible that he was a deep-sea diver and the watch was necessary to ensure he didn't get the 'bends', whatever they were.

Should one still desire to own such a monstrosity, the make of choice at the time was the Sekonda. This watch was made in the USSR and, in a sinister effort to undermine the free world, was sold in the West at a ridiculously low price. In fact you could buy one of these examples of soviet technology for about £5. It was popularly believed they were still made by slave labour somewhere in the Urals.

So there you have it; contrary to popular belief, we were a smartly turned out bunch with a keen eye for colour co-ordination, much given to understatement.

At least we thought we were, until one Sunday, Sid and I were invited to lunch on the tanker. This was the day when they had a traditional 'Rice Table'. The crew were Indonesian and they prepared a wonderful selection of Eastern food, with rice as the basis for each dish.

Not only was the food a superb change from our chip-based meals, but the table settings would have made the Savoy envious. Crisp white tablecloths with napkins to match, crystal glasses, silver cutlery and monogrammed condiment sets adorned the table. This in itself was manageable, until we noticed the disparity in dress between our hosts and ourselves. All the Officers were turned out in crisp white short-sleeved shirts with shoulder epaulets denoting rank, sharply creased black trousers and the shiniest shoes ever seen outside of a crematorium.

Now, being aware of the honour bestowed on us, Sid was tastefully dressed in his baggy cardigan with a hole in the left elbow, a pair of oily safety boots (slippers, he felt, would not be right for the occasion) and for some unaccountable reason, a scruffy tie with the left side shirt collar trapped underneath. Fortunately, to redress the balance, so to speak, I had on my best fleece jacket. Admittedly, you couldn't zip it up and the elasticated waist had expanded to about twice the size. However, my boots were reasonably clean.

After the meal, coffee was served in china cups with the name of the ship embossed in gold on the side. Sid and I were now well past being embarrassed and tucked in to sweetmeats of various kinds, with second and third helpings of coffee, all served discreetly by Kato-like waiters in white jackets.

Afterwards I was invited by the Chief Engineer to inspect the Engine Room, but first we went to his quarters so that he could change. Change in this context meant slipping on a pair of brilliant white overalls, complete with creases in the trouser section and his name and rank embroidered on the pocket.

Why his cabin was called 'quarters' I will never know. What he had was a stateroom with private bathroom, lounge and office, all *en suite*. His hobby was building model steam trains, one of which was in the process of construction on a huge table. The train was a work of art in gleaming copper and brass and he was rightly very proud of his endeavours. I hoped he wouldn't ask me if I had a hobby when off duty; there isn't that much room to build things on a scaffolding plank.

Just before we left to go back, Sid, in a moment of weakness, asked the Captain whether he and his Officers would care to come over to us for lunch, as we would like to reciprocate their hospitality.

All I could think was, 'Oh no!.............. Sid, what the bloody hell are you saying?'

Even if we rebuilt the platform, got rid of drilling, polished our stewards and demolished the recreation room, we still couldn't pretend it was anything other than a scruffy, oily, cramped metal box on legs.

Fortunately the First Officer answered with a polite refusal, saying how busy they were, etc. but maybe it would be possible one day in the future.

At that and with considerable relief on my part, we made our goodbyes and jumped into the waiting Bolkow.

When we got back, I said to Sid, 'Are you trying to reinforce the view throughout the tanker fleet that we are a bunch of scruffy Herberts? What the hell are you doing inviting them over to us? You know how bad our culinary conditions are compared to theirs and we haven't got a stick of mahogany panelling anywhere.' You could see I was in the grip of some strong emotions, mostly to do with snobbery, but one has one's standards after all.

Sid said, 'I know, it just came out, I felt we should show them that we can be hospitable as well, but I have to say, I am damn glad the Officer turned us down.'

I said 'That's probably because that same Officer has already been on board!' I explained that he had come to talk to me about mooring problems and I, in a moment of weakness, had invited him to lunch.

However, when we got to the Galley there was a queue stretching the length of the corridor and as my guest and I tried to walk to the front, we were threatened with our lives by a bunch of irate mud-covered drillers.

Much to my relief, on realising that he was about to be killed, the Officer said that on reflection he wasn't very hungry and he aught to be getting back to the ship.

'Bloody hell,' said Sid when I had finished, 'That could have been embarrassing, they would have been one officer short.' Sid's grasp of reality when faced with potential disaster was awesome.

I replied 'It wasn't so much the aggro I was worried about as the awful possibility that one of the drillers would get too close and splatter oil-based mud all over the Officer's gleaming white shirt.'

'You're right,' mused Sid, 'We would have had to offer to wash it for him and it would have been returned in a sort of shrunken off grey! It doesn't bear thinking about.'

17

Medical Matters

Think of a love match between the popular (but incorrect) image of Florence Nightingale and Klinger from MASH and you have a reasonably accurate picture of an offshore Medic. Ours was a hairy character about six feet two inches tall who hailed from Tyneside. This made consultation somewhat difficult, as most of the patients couldn't understand his questions and possibly more importantly, his diagnosis.

Offshore Medics had the authority to diagnose and treat patients within a set of largely unknown guidelines; this meant that adherence and interpretation depended mainly on the Medic's thirst for customers. A bit like Dr Knox, but the bodies were still warm.

Fortunately (for us) they rarely dealt with emergencies, but often had to care for people for some time before evacuation was available. The majority of cases were much more akin to a GP consultation, after which the patient was likely remain on the platform, a victim of the Medic's tender mercies —known in the trade as 'aftercare'. A word that often struck terror in the recipient.

The most common injury was a fracture or suspected fracture and the most common body part affected was the hand. This confirmed the age-old belief that offshore workers couldn't keep their hands to themselves.

The Medic's inner sanctum was the Sick Bay and the one area of the platform where cleanliness shone forth. Andy was our regular Medic and, in order to maintain the requisite hygiene levels, had trained a member of the catering staff in the specialist art of deck swabbing. This key task was undertaken daily by the appointed swabber, who was plainly overawed by the array of shiny instruments, skeleton diagrams and the smell of Lysol. Awareness of his frontline role in the need to combat germs ensured that he scrubbed and polished like a man possessed, even changing the water in his bucket without being told to do so. No problems with MRSA for us then, although I don't think it had been invented when we were offshore.

The Sick Bay was clearly identified with a big green cross on the door and was kept locked with the keys chained to Andy's belt. Obviously, the need to safeguard the film projector was paramount. For some strange reason, the Company believed they had to justify the cost of having a Medic on board and they felt this was difficult to do if he just sat around all day waiting hopefully for someone to become ill or injured.

So, in a stroke of mismanagement genius, some idiot thought it would be a good idea for the Medic to help out in the galley (carrying fruit and veg boxes, tidying up in the canteen, etc.)

Fortunately, Sid immediately cancelled this lunatic proposal and suggested that Andy could undertake routine health surveillance, monitor food and water hygiene and provide basic first aid training.

'However', said Sid on imparting the new regime to Andy, 'none of this should jeopardise your ability to show films on time.'

One of our less highly regarded 'stand-in' Medics was fond of reiterating in a lofty fashion, that secondary duties could not be allowed to disrupt his regime and he must always be immediately available to respond to emergencies. He said he couldn't undertake any of Sid's ancillary tasks because this would compromise his primary role as a Medic.

As emergencies were few and far between and some of Sid's so-called 'duties' required him to leave the Sick Bay, we were convinced he was just a lazy sod.

As Sid commented, 'I'm in a bit of a quandary over giving the miserable git other jobs to do, it's a bit like insulting a Chef—they can extract a terrible revenge without you even knowing it's them.'

Fortunately, the guy only made a couple of trips with us and when I asked Sid what had happened to him, he shrugged and muttered something about 'incompatibility'. This was Sid's code for 'I'm not having that bastard on board again,' and was a sure sign that he had been up to his usual 'behind the scenes' lobbying.

(This unofficial approach to hiring and firing later became known as 'NRB' and was shorthand for 'Not Required Back' which a Supervisor or Manager would append to someone's timesheet as they left the platform. It worked quite well for a time until the Trade Unions got wind of the code and began to cry 'unfair dismissal'. Your moral view depended to a large extent on whether, in using the system, you had managed to quietly get rid of one or two useless toe-rags. I of course, couldn't possibly comment.)

Fortunately, Andy had a much more relaxed attitude to work and was often seen helping out when the sick bay was closed. As he said, sitting there on his own and waiting for a chance to show off his medical skills, he was reduced to playing tiddlywinks with the Paracetomol and marbles with the laxatives. Often the highlight of his day was to dispense Lemsip, cough drops or the odd suppository. He once confided that frustration on this scale made him long for the good old days when poor diet meant he had lots of lovely boils to lance.

We, on the contrary, felt that our continued good health was a guarantee that Andy's key role was never disrupted. I refer of course, to his custodianship of our films, the Projector, Splicing gear and Re-winding machine. All of which were lovingly cared for by Andy.

Such was Andy's awareness of the importance of recreation to our well-being that he also maintained an unofficial lending library. This consisted of a motley collection of dog-eared paperbacks, but which nevertheless could be a life saver if you ran out of something to read. He started the collection by simply purloining any carelessly discarded books left in the recreation area. He even had the stewards looking out for any stray books, but this caused some havoc when it was discovered that one steward, anxious to please the Medic, had started to take books from bedside tables when cleaning the cabins.

Surprisingly, this fine example of individual enterprise led to some distress and a number of attempts to kill the steward. As Andy said, 'Injured stewards I can deal with, but dead ones are beyond me.' You don't get care like that from your average GP.

The library system was simple but strict; to obtain a book you had first of all to offer one in exchange. This meant there was no need for complicated tickets or issue and return stamps.

Sadly, for many would be users of the service, no book in meant no book out and anyone who suddenly returned with a book for exchange was subjected to close questioning with regard to its provenance.

Offers to bring out two books next time also received short shrift from Andy. There is no doubt the British Library would have been overawed by the simple elegance of his system - but not for long. That may be because it often involved physical threats to the person, something that most librarians are not exactly noted for. (Shushing and glaring, yes; violence, no.)

In between the important work of filming and the discharge of library books, there was still time for Andy to offer medical consultation. He had a fund of stories regarding the average male's approach to issues of a medical nature and we had many interesting chats late at night, while taking a mug of restorative tea in the sickbay.

I recall him saying that one of his favourite opening gambits came from a contractor who, on entering the sick-bay, said, 'I don't want a bloody diagnosis or anything, just tell me what's wrong and what can you give me to fix it?'

I noticed during one of our off duty sessions that Liam seemed to lack enthusiasm for Andy's tales. When I asked him why, he said he found it hard to trust a Medic who kept a dead plant on his shelf.

But the best story of all was one he told about the trapped foreskin. We all leaned forward to listen.

One evening there had been a discreet knock on the Sick Bay door, accompanied by what sounded like a sort of low sobbing noise. On opening the door, Andy was confronted by the sight of a guy bent nearly double and clutching his private parts. But the most surreal part, said Andy, 'Was the fact that he was wearing a survival suit.'

Letting out another low moan, the chap shuffled slowly into the room, which took quite a while, due to the fact that he was taking minute, very careful steps.

Andy, still puzzled but fascinated, said, 'Sit down and let's see what's wrong.'

However, the chap, who was now leaning against the work top, said, 'I can't bloody sit down, I'm in agony and it's taken me nearly an hour to get to you.'

To which Andy replied, 'Well if you don't let me know what's happened, I can't help you.'

Realising he would have to explain, the chap told his story.

Apparently, he had undressed and had been just about to go to bed when he had what seemed to be a good idea. As he was going ashore the following day, he thought he would try on his survival suit to make sure he could get into it without any delay.

All went well until he came to pull up the zip, giving it a good heave as he had been instructed. The zip shot up and within a nano-second our man experienced a searing pain the like of which he didn't believe was possible. The zip had caught the end of his foreskin in its teeth. As any movement on his part increased the agony, he couldn't bend forward sufficiently to see the extent of the damage and he wasn't about to try and pull the zip down again.

As we listened in rapt silence to this terrible story, Liam whispered 'Bloody hell, it brings tears to the eyes just thinking about it.'

Andy nodded and continued with the saga. He said to the victim, 'Look I'll have to move you onto the bed so that I can get at the zip and see what the damage is.'

A word here about the zip. Don't be fooled into thinking, 'Oh yes, I know what you mean, a big zip just like the one on my fleece jacket.'

No it wasn't—the zip in question was frightening, it was huge, it was made out of armoured steel from old World War 2 battleships, it had teeth like an alligator and was enclosed in a tight rubber seal on both sides. It took all your strength just to pull the damn thing up, never mind trying to pull it down again. Comparing this monster to any other zip is like comparing Frankenstein's Monster to Mary Poppins.

Anyway, on hearing this medical examination proposal, the guy stared at Andy with mounting horror—recognising he couldn't stay as he was, but terribly afraid that our Medic would grab hold of things with his size twelve hands and start ripping them apart.

Realising he had no option, however, he shuffled towards the bed and with renewed sobs and curses managed to climb on, while at the same time never relinquishing his grip on the source of the pain.

As Andy started to put on a pair of rubber gloves and began arranging an assortment of tools on his tray, the patient stared at the shining heap of metal and whimpered, 'What the bloody hell are you going to do with that lot?'

'I'm going to see if I can release your prepuce from the zip.'

At this point, Bert interrupted, 'Hang on Andy; I thought you said his foreskin was trapped. What's this prepuce thing?'

Andy gave one of those superior looks so beloved of accountants, and said, 'Prepuce is the medical name for foreskin. I thought everyone knew that.'

'Never mind all that medical crap', snapped Bert, 'Just stick with foreskin. Now—what happened next?'

'Well,' continued Andy, unruffled, 'Once the guy was lying on the bed, I had a closer look. I was able to put my hand inside the suit and could feel where he was trapped by the zip.'

'Was it bleeding?' asked Liam, who has a macabre thirst for such details.

'Of course it was bleeding! When I removed my hand the sight of blood on my glove did nothing to reassure the guy either—he nearly fainted. Anyway, I could see no alternative but to gently start to pull the zip down whilst at the same time holding onto his willy.'

I looked at the lads and couldn't help noticing we were all surreptitiously pulling our jeans away from our own willies and wriggling slightly in our seats, the better to be reassured that all was well in the trouser department. I suppose it was an example of the power of positive thought.

Then, revealing a wholly inadequate bedside manner, Andy had said to the quivering patient, 'This may hurt a little, but I need to pull the zip down, just hold on while I make a start.'

'Bloody hell,' said Liam, which just about summed up our feelings for the poor sod at that moment.

'Anyway,' said Andy, 'By slowly moving the zip down and at the same time holding onto the guy's affected parts, I managed to free him from the clutches of the teeth. Of course, the exercise was accompanied by the guy shouting ineffectual instructions like 'Hold it there, stop pulling, oh my god, I'm going to kill you; and they can shove this job - I'm never coming offshore again.' But at last I got him free, thought there were some wicked looking cuts and abrasions to the aforesaid foreskin and quite a considerable amount of blood. I have to say it wasn't a pretty sight, almost as though someone had tried to circumcise him with a potato peeler.'

Liam winced. 'Listen Andy, I know you're enjoying this, but just leave out the gory details and tell us what we all want to know—was it all over for him as far as marital duties are concerned?'

'Funny you should ask', said Andy, 'That's just what our friend wanted to know, once I had cleaned him up.'

'Well', said Liam, 'What did you tell him?'

It was at about this time we began to think that Liam was taking things a little too personally, but we all wanted to know anyway.

'I told him it would heal up in time, but to remember that the foreskin contains specialised blood vessels and nerve endings and that in some cases there may be a need to carry out prepuce reconstruction.'

Liam again, 'What the hell do you mean by *reconstruction?*'

'It means that the foreskin might have to be stretched in a non-surgical procedure in order to ensure adequate coverage.'

'And how do they do that?'

'Well one method is to hang weights on the end until it stretches sufficiently.'

Bert had heard enough. 'Sod off Andy; you're making me feel ill. Just tell us what you did to finish the job.'

'Well I told the guy I would have to put a dressing on it, but as I couldn't use a bandage, I fixed the dressing in place with plasters.' He looked meaningfully at us.

'The guy then asked me if it would heal quickly and whether there would be any permanent damage. I reassured him that the damage was fairly superficial and things would be okay in time. But one precaution he should take was to forget about Sophia Loren and definitely not to get too excited about marital duties for a while.'

There was a respectful silence for a time as we digested the information and tried not to visualise ourselves in that position.

Then Liam asked one final question. 'How could the guy get the plasters off his willy without causing himself even more pain?'

'Exactly,' chuckled Andy and we made another well-earned brew.

18

Away all Boats

Although the title is not strictly relevant to the following, it sets the right dramatic tone. I have liked the phrase ever since I saw a film of the same name. It was a Second World War drama starring Jeff Chandler as a Captain of an Attack Transport Vessel, a man with the kind of 'gung ho' heroism rarely seen, until that is, we entered the fray.

You will see that what follows entirely supports my contention and please, don't pretend you've never heard of Jeff Chandler.

Anyone arriving on the platform for the first time was given an 'Induction' briefing by one of the regular staff. Its purpose was to explain what they should do in the event of an alarm sounding, with an emphasis on the need to obey orders during an emergency. The presenter worked through the Station Bill, which set out basic rules to be followed, and the exact location of the individual's muster point and lifeboat station.

Lifeboats were located at various points along the bottom walkway and in an effort to make things clear, we used to explain that the accommodation was at the north end of the platform and the flare boom was at the south end. On the occasion of which I speak, the listeners had just about grasped these complex facts, when Liam, who regularly acted as a presenter, asked a guy in the audience if he knew where his lifeboat was. After staring at the Station Bill for an eternity, he said he thought that for some strange reason it was the one marked as 'North-east.'

'Correct,' said Liam.

Confirmation that he had cracked the code somehow failed to cheer our newcomer. When Liam asked what the problem was, he replied, 'I don't know where the northeast *is*. You said we had north, south, east and west, you never mentioned anything about sodding north-east.'

Liam, in an effort to reassure, replied 'Look, it's no problem, think of the platform as the UK; you live in the north-east of the country.'

'No I bloody don't,' said the confused one, 'I live in Newcastle.'

Liam, realising he was on a loser and bluffing like a politician, just said, 'Look, don't worry—follow the arrows on your Station Bill and they'll take you straight there.'

Unfortunately, as usual what happened was, when the signal to go to the lifeboats was sounded, everyone left their respective muster stations, which were located in the accommodation and started to move *en masse* down the outside stairs.

Our friend from Newcastle could be seen staring at his Station Bill and muttering 'north-east' to himself as he set off.

Sadly, once on the stairs, he found he couldn't see the flare boom and as far as he was concerned the stairs faced south, which was obviously the wrong direction. Things became even more fraught when, on reaching the first landing, he had to turn west then descend the next flight facing north. The next landing forced him to turn east before the final flight sent him south.

By then of course, he had absolutely no idea in which direction to go in order to get to his lifeboat, and his Station Bill had been turned over so many times that the arrows were now, as far as he was concerned, meaningless.

However, on turning the corner of the platform, a minor miracle occurred, that is as far as he was concerned, and he suddenly came across a lifeboat station full of people putting on lifejackets.

'Go no further,' he thought, 'This'll do for me.' Grabbing a lifejacket, he joined the queue. Sadly, as he was about to find out, he was in the wrong place and consequently wasn't on the list for that boat. This was brought home rather brutally when the muster-checker for that boat said, '*You're* not on *my* list pal, so sod off and find your own boat.'

It was amazing just how many times this lost passenger situation occurred, resulting in tannoy broadcasts having to be made in an attempt to locate the missing. It was like a giant game of blind man's bluff, with lost people being re-directed all over the platform. How we ever managed to account for everyone is still a mystery—but we did.

Anyway, once he was sure that everyone fully understood the muster and lifeboat procedure, Liam had then to demonstrate the second and perhaps the most dreaded part of the induction —the use of the Lifejacket.

Lifejackets had to be put on before entering the lifeboat. They were stored in huge metal boxes adjacent to each lifeboat station and on arrival each man picked one out of the box and swiftly and efficiently put it on in readiness for entry into the boat.

At least in theory.

Our lifejackets should not be confused with the silly little things you see demonstrated by smiling Airline cabin crew. Ours were huge unwieldy efforts shaped like a tabard, which you pulled over your head. Then you had to slip your arms through the loops of two long tapes. The idea was you then grasped the tapes and pulled them firmly downwards through plastic rings to tighten the open sides. Then, you wrapped the tapes, which were about four feet long, around the jacket and fastened them across your chest.

Describing the method on paper is bad enough, doing it for real took you into the realms of lunacy.

Buoyancy, in our lifejackets, was achieved in mammoth style by having a number of brick shaped pieces of hard polystyrene sewn into the front lining. These reached from just under the chin to about the waist, depending on how tall you were. When the jacket was fastened, one's ability to move and, in particular, bend, was severely restricted. It could also be downright dangerous; if the wearer dropped something and bent down to pick it up, the polystyrene blocks would ride up and deal him a smart blow under the chin. All this meant that effecting a smooth and dignified entry into the lifeboat was almost impossible.

Generally, at the briefing the presenter, having practised for weeks in secret, would demonstrate the correct way to wear the jacket and then get a 'volunteer' to have a go. This invariably resulted in much hilarity from the assembled group and considerable embarrassment for the volunteer.

Invariably the tapes became tangled under and around his arms and neck. He would struggle to find the ends, which by then were inextricably wound around the inside of the jacket. The upshot was, he couldn't fasten the tapes properly around his chest and would typically be left with one tape missing and the other tape dangling around his knees.

It was one of those situations where, in order to avoid further embarrassment, everyone assured the presenter they had learned from this experience and wouldn't fall into the same trap as the unfortunate volunteer. 'There's no need to worry,' they would claim to a man, 'We are confident that putting on a lifejacket is a simple task and could be accomplished blindfold.'

Let's just say that from bitter experience, we knew this to be rubbish and I often thought that fitting our lifejackets would have made an excellent challenge for Bruce Forsyth in the Generation Game.

Later on, when I began to travel regularly by air, I used to enjoy watching the gorgeous lady in her tailored suit demonstrate the lifejacket before take-off. And each time the simplicity of the task would be emphasised. The punch line was always the same; she would finish with the deceptively simple instruction to 'Tie the tapes securely at the side with a double bow.'

I often thought that not one half of those on board would have any idea how to translate her instructions into successful actions, and I bet she knew that as well.

Enough of the philosophising, it's time to explain what happens next. Remember the incredibly clever title?

Well, our lifeboats hung from davits over the edge of the walkway in readiness for an emergency departure. Just what might happen to make the entire crew take to the boats, no one at the time cared to speculate. I suppose that was naivety born of ignorance.

They were bright orange and totally enclosed with an entrance hatch in the centre and a porthole in the top, so that the Captain could see where we were going—or not, as the case may be. The other laugh was that each boat was 'designed' to carry up to fifty persons.

Power was provided by a diesel engine, which was started by pumping up a pneumatic cylinder and then releasing the compressed air to turn over the engine. Having done so, you then had to close the door and pull on a wire rope threaded through a hole in the roof. The fixed end was attached to a winch motor located at the side of the davits. Pulling on the wire started the winch, which then unwound two lengths of wire rope, which were attached to mountings fore and aft of the boat. This gently lowered the boat onto the sea. Unfortunately, this part of the procedure had been written by the same lunatics who thought the boat would hold up to fifty persons.

Anyway, once afloat, the Coxswain (Liam or Bert, who else?) pulled another lever and a quick release mechanism freed the boat from the ropes, after which he simply motored away from the platform.

Personnel could be evacuated from the life-boat in one of three ways: by transfer to an inflatable fast rescue craft, by winching up into a search and rescue helicopter, or by transfer to a larger ship such as a stand-by vessel.

Again, remembering our own rescue efforts, this was the theory.

Each week, usually on Sunday, we held an emergency drill to ensure that everyone on board knew how to get to his or her designated muster point and lifeboat station. At a given time, the platform alarm would sound and everyone made their way to the check-in, where the 'muster-checker' carried out a head count and compared the results to the 'Persons on Board' list.

When the OIM was satisfied that everyone had been accounted for, the alarm status was changed to 'Abandon Platform' and we all made our way to the lifeboats.

As Sid, in another fit of management, had decided I should be the Captain of one boat, it seemed only fair that I appoint Liam to be the Coxswain.

My job was to tick off each person arriving, after which I guided them into the boat and they sat down and fastened their safety belts. In the meantime, Liam would open the top porthole, prepare the release mechanism and prime the engine starter. When all my passengers were 'ticked off', I notified Sid, climbed aboard and fastened the watertight door. At this point, Liam would start the engine and we were ready to launch.

The exercise was complete when the engine was running and everyone on board accounted for, after which we all went back to the accommodation to brew up and watch a film.

In practise, a number of things tended to reduce the enjoyment to be had from an exercise well done. These 'things' tended to fall into two variants, the inanimate and the human.

The method of priming the engine starter was medieval, to say the least. A lot of air had to be compressed and this was accomplished by means of a three-foot steel bar inserted in a slot in the accumulator and pulled back and forth for fifty strokes. Now, this sounds easy, until you consider the restricted room in the lifeboat and the fact that with every stroke, resistance to pulling and pushing increased. Meanwhile, as this epic struggle was taking place, people jammed into unbendable lifejackets were falling into the boat, trying to find seats. Liam would be lathered in sweat, cursing the clumsiness of the passengers and trying not to lose count of his strokes.

I mentioned earlier that the capacity of the boat was claimed to be 'up to fifty persons'. This was obviously a clever marketing wheeze on the part of the manufacturer. It must have been based on the assumption that the passengers were about four feet tall, weighed no more than six stones and were naked. In fact, by the time the first ten or twelve passengers had clambered aboard, found a seat on the bench, and strapped themselves in, the boat was full.

This was a problem, as I would still have at least another dozen guys queuing on the walkway, in the fond belief there was ample room inside for *them*. It is also important to remember that during these exercises we all wore overalls, warm clothing, lifejackets, and safety helmets.

In order to break the log jam I had to lean in through the hatch, pretend to be Captain Bligh and yell profanities at the incumbents to make them move closer to one another. This strategy required a delicate balance of bad language. In other words I could swear enough to frighten them, but not enough to cause a minor mutiny. Managing overcrowding on the London Underground would be child's play when compared to filling the average lifeboat.

Finally, exhausted, we were all aboard and I would give Liam the signal to start the engine. Sadly, more often than not the engine would briefly turn over, splutter a bit and *not* start. This meant that Liam would have to prime the accumulator all over again, in a packed boat with thirty pairs of eyes all staring at him, wondering if this would happen in a *real* emergency.

It is at times like this that the true mettle of a man is revealed and rising to the challenge, Liam usually did the only thing possible in such circumstances, throw the steel bar at one of the passengers and say, 'You bloody well have a go, I'm knackered.'

Surprisingly, this tactic rarely succeeded and he would be given back the priming bar with succinct instruction regarding where, anatomically speaking, he could put it.

On these occasions, if I could manage to straddle the engine, I would give him a hand to achieve the magic fifty strokes. Usually, quite a hush descended on the passengers during this time, as the realisation began to dawn that failure to start the engine could pose a problem if, or when, we were in the water.

Fortunately having turned over at the first attempt, the engine usually fired the second time and Liam could fall back exhausted. I do recall one occasion when one of the passengers, being new to this situation and displaying a rather nervous disposition, gave hesitant cheer. I will always cherish Liam's look of astonishment and his teeth-gritted query as to whether his newfound admirer was 'taking the piss'.

Once we were all safely in the boat, the engine running and Liam ready with the winch engagement wire, I awaited the next command. Unfortunately the longer we waited, the noisier the engine became and the more the boat filled with diesel fumes. This was when the passengers began to get restless as they realised there was potentially a stark choice between drowning or slow asphyxiation from lack of oxygen.

I would begin to pray that Sid would call the exercise over and we could be released from the ever-deteriorating situation. Glancing at Liam, there was often a look on his face that indicated he secretly enjoyed the growing discomfort of the passengers. I think he viewed it as a sort of revenge for their ingratitude over his exertions on their behalf.

A year or two later, I was once again the captain of a lifeboat on one of the northerly concrete platforms, but this time there were two significant differences. The people designated for my boat were the drilling crew, which might not have been a problem in itself, except that they were all built like brick bus shelters and German.

Don't misunderstand me, their nationality was not an issue—what *was* an issue was their names.

The first time we arranged a lifeboat drill, I took my check-in list and went to the boat to await the arrival of my passengers. As they gathered, I made ready to call out their names and that's when the problem started. I stared at the names on my list and realised with growing embarrassment (or was it fear?) that I couldn't pronounce them. However, looking up at the sea of expectant faces, I made a start:

Hans-Jurgen Vollbracht
Cramon Leinsberger
Pfugbeil
Seikemuss

Trubenbach
Aschenbrenner
Chamier-Glisczinski

There seemed to be a growing pause between my calling the name and hearing a reassuring *'Ja'* from the correct individual. As I ploughed on, I began to hear Germanic chuckles followed shortly by the less reassuring sound of muttering amongst the crew. I could see this wasn't going to work and it became obvious that my phonetic pronunciation was causing a mixture of puzzlement and hilarity among my Teutonic friends.

Fortunately, I was already acquainted with their Maintenance Supervisor, a red haired giant named Helmut (the only name I could pronounce.) He had been most helpful to me one midsummer's evening when I had to renew a huge caisson fixed to the outside of a concrete leg. ('Caisson' is just me showing off again, it's really just a thirty-six inch diameter pipe down which we hung a submersible fire pump.) We had successfully installed it by the light of the midnight sun and as a result, we became good friends.

Seeing Helmut in the group, I grabbed him and said, 'Helmut, if we're ever going to get your lot into this blasted boat, I need you to call out the names and I'll tick them off—otherwise we'll be here all night.'

Fortunately the new system worked and they began to get into the boat. Then with a distinct sense of 'deja-vu', I could see congestion building around the entry hatch. Having learned the hard way, I pushed my way onto the boat began physically shoving the incumbents to the furthest seats, never, of course forgetting to keep up a steady stream of apologies; it was not for me to undo good Anglo-German relationships.

This time I was in no doubt they understood me and we managed to get about twenty-five big hairy Germans most of whom were still bemused at my efforts to pronounce their names, all seated. Even more fortunate as far as I was concerned was that at the time they were working with a 'skeleton' crew. I shudder to think what might have happened if they'd turned up with a full complement.

One other embarrassing incident regarding the Germans sticks in my mind. Their Toolpusher was a very dignified chap by the name of Karl and as he and I used to attend management meetings together, we got to know one another quite well.

Once, as we were making room on deck for a new gas compression module, I offered Karl a steel staircase which we were taking down, as it was in our way.

He readily accepted my offer and within about two hours the stairs were dismantled, swung over the helideck, re-sited and welded in place on his drilling rig.

Later that afternoon as we were chatting about the refurbished stairs, the Medic announced he had acquired a 'must see' film. So later that evening I found myself sitting next to Karl as we waited for the mystery film to start. The lights went down, there was a hush of anticipation and there on the screen accompanied by the London Symphony Orchestra at full blast, were the immortal words, 'Sink the Bismarck.'

Time to go back to the lifeboats.

In the early days before common sense prevailed, we used to launch a lifeboat and have a sail around the platform. This idiotic enterprise usually took place on Sunday mornings when the sea was calm, under the guise of 'necessary maintenance.' We, that is Liam, Bert and I, would open up the boat, prime the engine, release the winch lock and descend nice and slowly into the sea.

Oh how people envied us; there we were in the sunshine with our heads peering out of the top hatch or standing on deck, casually holding on to the grab rail. However, being kindly souls and in a fine spirit of camaraderie, we often used to invite one or two people to accompany us on our jaunt. We got into the habit of asking members of the catering staff, as they seldom left the accommodation and knew very little about the workings of the platform.

The fact that this noble gesture on our part would ensure extra towels in our rooms and special sticky cakes delivered to our office had nothing to do with our choice of passengers at all.

One person we used to regularly invite was Sid, who just as regularly refused, saying he had been in enough silly situations to last him a lifetime. This pearl of wisdom would come back to haunt me a year or two later in a rather forceful manner. However at the time we would enjoy our nice little sail and as there were usually only about four of us on board we used to leave the hatches open and encourage our catering friends to take in the fresh air and the view, especially the enormous size and height of the platform when seen from sea level.

Each platform had a 'Standby Boat' whose task it was to hover by in case the crew had to abandon the platform. The theory was that we would sail our lifeboats to the standby boat and then be taken on board to safety. The only downside to this was the fact that they were old redundant fishing boats, liberally covered in rust, and saved from the scrapyard by the needs of we 'offshore tigers'. Watching them from a height as they slowly circled the platform was reassuring if the weather was fine; however, watching them in bad weather was somewhat disconcerting as they would completely disappear in the trough of a wave only to reappear a minute or two later, bobbing about like demented corks.

One morning we had a call from our stand-by boat asking if we had any food to spare. They were contracted to stay on station for a month at a time before sailing back to base for a crew change and re-stocking. However, this time there was some problem with their replacement and they had been told to stay with us for longer.

Sid asked us to take some food for them across and as the helicopter couldn't land on a cork, we were to go by lifeboat. Being still in idiot mode, this was all the excuse we needed and off we went clutching bags of potatoes, cabbage, carrots, bread and assorted tins of food. We felt quite noble riding to the rescue of our erstwhile rescuers; I know it's childish, but you should have already realised this by now.

On approaching the vessel we were thrown a rope, which we made fast to the cleats on our boat. A short (thankfully) rope ladder was hung over and we clambered on board. We passed the victuals to them and to show their gratitude, the Skipper asked if we would like to see his facilities before returning. Being curious and mindful that this was better than work, we agreed.

The vessel seemed to be very small compared to our platform and was typical of that old type of fishing boat. There was rusty steel everywhere and the galley would have fitted comfortably in a phone box. However, the part of the vessel he was most anxious we should see was the area under the deck. This space had been cleared of fishing equipment and in its place they had fitted rough wooden benches around the perimeter walls. This he told us proudly was where we would all be located if we had to be rescued from the deep.

We looked at one another and trying desperately not to appear ungrateful, I managed to say how good it looked and how sure I was that we would be in good hands. Liam said very little, except to indicate that we should be getting back, so we thanked the skipper once more and beat a dignified but hasty retreat.

We got back on board and reported to Sid. 'Well,' he said, 'What did you think of the fishing boat?'

I muttered something along the lines that they were a lovely bunch of people, but couldn't understand how anyone could spend so long just circling around us in that way.

Liam said, 'I just can't believe we could all fit into that small space, it would be full with just the drilling crew on board.'

Sid, ever reassuring, said, 'Well you know my view; the best way off this place in an emergency is from up there.'

He raised his eyes toward the helideck. We all nodded in some relief and remembered that this method had been highly successful in evacuating everyone after the supply boat had taken a short cut through our jacket. Once again, the oracle had spoken, all was well and a nice cup of tea seemed to be in order. Looking back, I now realise how lucky we were at the time.

One evening just as we were finishing dinner a call came over the tannoy. *'Man overboard!'* This we had dreaded, as the chance of finding someone in the North Sea in poor light was remote to say the least. We set our emergency procedures in progress and asked both the standby and supply boats to come gently towards us and turn on their bridge lights. In the meantime, people lined the railings, holding onto life-rings whilst peering hopefully into the sea.

Liam, Bert and I went under the deck onto a low walkway, but we could see no sign of anybody in the sea between the platform legs.

Sid, who was co-coordinating the search, was in touch with us all by radio while trying to find out who had given the alarm and to get some accurate information regarding where, when and how the person had gone overboard.

We continued to search with a growing sense of concern for the unfortunate victim, but the light was rapidly failing. The two boats were running a box search, but to no avail.

Then, we heard another tannoy message from Sid calling off the search and assuring us that everyone had been accounted for.

Somewhat puzzled but very relieved, we made our way back to his office to find out what had happened. When we arrived, he said, 'Shut the door.' I noticed that the Drilling Contractor's Toolpusher was also present.

Sid said, 'We've found out who broadcast the message, it was one of the drilling crew.' We were astonished at this news and asked what was the hell was going on.

The Toolpusher, staring at a space some miles away replied, 'The bloody lunatic says he did it for a joke and didn't think anyone would take any notice.'

Liam stared at him. 'Didn't the idiot realise he could have placed many other people in danger? Not to mention having to call up two boats for a search in near darkness.'

I will always remember the Toolpushers face as he turned to Liam and said, 'Oh yes, he understands only too well the goddamn problems he's caused. Tomorrow he gets his ass fired off this rig and he will never work for any other drilling company in the goddamn world again.'

I could see he was seething with a mixture of embarrassment and anger, so in order as I thought, to defuse the situation, I said, 'Thanks for sorting out who it was, where is he now?'

This turned out to be the wrong question.

The Toolpusher turned to me and growled, 'Never you mind where the lunatic is. Let's just say he won't be eating any supper and if he as much as squeaks, I'll swing for the bastard.'

I thought, wow, yet another valuable lesson in management by direct action.

My expeditions in lifeboat seamanship came to a sudden, dramatic and embarrassing conclusion some time later, on one of the biggest concrete structures in the North Sea.

As was by then our custom, we decided to launch a boat and have a nice sail around the platform. There were two maintenance mechanics with me and this time we had invited the Safety Officer and a member of the catering staff to accompany us.

So down we went and off we sailed. Two things then occurred which had a marked effect on our enjoyment; the first was that the engine stopped and the second was that the sea was not quite as calm as it had looked from eighty feet above.

Without an engine the boat began to wallow from side to side as the swell took control, this resulted in the craft adopting a rather uncomfortable motion. Very soon, both our passengers were seasick and had completely lost interest in the voyage. The mechanics and I frantically pumped the starter, but the engine still refused to fire. It looked as though we had a blockage of some sort in the diesel system. However as we had no tools with us, there was little we could do except keep pumping.

In the meantime we were at the mercy of the sea and we began to drift alarmingly towards one of the concrete legs, which meant we were now virtually out of sight under the platform. I had no option but to call our standby boat and ask for their assistance immediately.

Fortunately, this standby boat was a purpose built vessel with a fast rescue Zodiac mounted on davits over her stern. They quickly launched it and came towards us at a rapid rate of knots. The swell was now in complete control of the lifeboat and we needed to get out from under the platform as soon as possible. The Zodiac came alongside and threw us a line, but we missed it. They swung around and made another slow approach towards us but much to our horror, a large swell lifted the Zodiac and bounced it hard against the concrete leg. As we watched, the whole side of the inflatable deformed and then bounced back. With great presence of mind the pilot accelerated away from the hazard and made one more run towards us. This time we made no mistake, grabbed the line and made it fast through the forward cleat. We were then towed slowly towards the winch lines hanging from the davits and again, we failed to grab them both on the first attempt.

This task was extremely hairy as two of us had to position ourselves fore and aft and hang on to safety rail located on the upper side of the lifeboat. This meant we had only one hand with which to catch hold of the wire rope and, to prevent us being pulled off the boat, the Zodiac had to reduce speed to almost a standstill. However, we did succeed and much to everyone's relief, managed to hook the ropes into the cleats. The Zodiac moved away but stood by until we were lifted clear of the water. Meanwhile I radioed the onboard mechanic to start winching us in, and that's where the second part of our little adventure began.

The winch began to pull the ropes tight, but just as we were almost out of the water; another swell caught us from the stern. Consequently, as we lifted clear we began to swing from side to side. At first this was fairly gentle, but as we were slowly winched up, the boat started to act like a giant pendulum. It was at about this point that I began to wish Galileo had paid more attention to the sermon instead of the swinging chandelier in Pisa cathedral.

As we swung ever more quickly from side to side, all I could think of was that we were a mass of some three tonnes hanging on the end of eighty foot long ropes. I remembered that Galileo had proved two things, both of which meant we were still in trouble.

The first was he observed that the bobs of pendulums nearly return to their release height, and the second was that changing the length of the rope would increase the speed of each complete swing.

Isn't it funny that when you should be thinking about serious things, such as drowning, you start remembering useless information about a long dead Italian? You would also have thought that his having invented the telescope would be enough for one man, but oh no, he had to worry me about pendulums as well.

Do you remember I mentioned our two seasick passengers? Well, suffice it to say that the new and exciting experience of being flung ever more violently from side to side was doing nothing to cure their malady. They were now inside the cabin clutching onto both the engine cover and their stomachs, and possibly regretting their decision to go for a sail.

We three 'experts' were also having difficulty in keeping on our feet, and so, pretending I was still in full command, radioed the mechanic and told him to stop the winch. This left us hanging about forty feet below the platform swinging, as predicted by Galileo, at a steadily increasing speed. After some minutes it became obvious it would take ages for the motion to stop, so I said, 'Sod this, start the winch again.'

Up we went, with the oscillations becoming shorter but faster until, with some alarming thumps we were at deck level. Our passengers fell out onto the landing and without a word of thanks, disappeared into the accommodation. We three then staggered out to a great cheer and raucous laughter.

What we hadn't realised was that word of our minor problem had been passed around the platform and the sight of both the Platform Engineer and two maintenance technicians being unable to start a lifeboat engine had been too good to miss.

As a result, the rails had been lined with people all there to enjoy the fun. I found it difficult to maintain any semblance of dignity at this time and my attempts to convince anyone that we really were in full control throughout were met with some cynicism.

However, the ultimate embarrassment was still to come as later that evening I found some budding poet had written a tribute to me and worst of all, had stuck copies on notice boards all over the platform. For some masochistic reason I kept a copy and this is what it said:

Come In Number 9

A lifeboat is dropped into the sea
Number nine it seems to be
The Platform Engineer's aboard this one
With the Safety Officer cabined in.
It leaves the platform far behind
Carves its course and as it winds
Around the platform it is steered
With all on board in good cheer.
This is changed when the engine stops
As no power is passed to the props
Radio calls are loud and clear
From the Platform Engineer.
We're drifting in this big North Sea
Could you get some help to me?
Arrangements made with the safety ship
The rescue boat soon makes the trip
To number nine lifeboat she does steer
And tows her to the lifting gear.
The lifting gear's attached to same
Number nine lifeboat's hauled up again
Secured and fixed in the Davit frame.
All come out of Number Nine
But two appear to take their time
Not surprising when they appear
They are greeted with a cheer
It's the Safety Officer and the Platform Engineer.

Now, you may agree that the poem may not be Wordsworth, but I'll bet it was received with considerably greater enthusiasm than his 'host of golden daffodils' thingy.

Suffice to say, it took me many days to live down the debacle, even though I tried to maintain what is known as a 'low profile'. There is no one like your peers for dispelling any grandiose maritime illusions you might hold.

This was the last time lifeboats were launched from our Installations and in the light of our experience rightly so. It also reinforced the view that lifeboats were designed to leave a platform, not to return to it.

Oh, and finally, just to salvage some pride, we did find a blockage in the diesel supply to the engine, but you try telling *that* to anyone.

19

Food for Thought

Mark Twain once said 'Part of the secret of success in life is to eat what you like and let the food fight it out inside.'

We certainly subscribed to those sentiments. At first, there was a certain novelty to offshore victuals. It wasn't just the availability—four sittings each day, tea breaks with fresh cakes, coffee, and fruit juice on tap. No; rather it was the sheer volume of food to be had, and what's more, all free. You could have second and even third helpings without criticism; if you timed things correctly, you could enjoy enormous cooked meals about every six hours. And custard came in a giant tureen.

The other untold joy was of course, the absence of wives or mothers ('partners' at that time were people you danced with—as in 'take your partners for a Quickstep';) who would normally issue sensible admonishments such as 'I think you've had enough,' or, 'If you eat all that you won't eat your tea,' or, 'I was going to make soup with the rest of that tomorrow!' or 'Must you pile up your plate like that?'

You know what I mean; it's a bloke thing.

Most people, on seeing the volume and spread on offer, thought they had died and gone to heaven.

In addition, your senses were under constant attack; for instance, there was the smell—fresh bread baked at night, bacon in the morning, steak at mid-day and dinner-time, all odours carried aloft through the extract fans to permeate the entire platform.

Good plentiful food was important for providing a high calorie diet for the men on the drill floor and others who worked outside in all weathers. For the rest of us it represented an excuse to enjoy a chat with colleagues in a wonderful chip laden atmosphere. Mind you when we arrived from the Buoy, frozen and hungry, we had no doubt in our minds that a substantial three or even four course meal represented the high point of the day and a just reward for our efforts.

Our catering staff were, in the vast majority of cases, very good. Many of the cooks had been recruited from the army or the merchant fleet and were well used to providing varied and substantial meals. You will notice that we didn't call them anything pretentious like 'chef'; they were happy to be referred to as 'cooks' and were proud of the food they turned out for us four times a day.

Once, while I was on another platform, it became obvious that sartorial standards had slipped to an unacceptable degree. The Camp Boss seemed unaware that his galley staff looked like a bunch of unwashed ditch diggers and the two cooks wore tee shirts that had once been white but now resembled Bruce Willis's vest just before the end of 'Die Hard'. Even the washer-ups looked like extras from a Charles Dickens adaptation, complete with facial pustules and lank hair resembling a badly treated floor mop.

Although the food was palatable, it was served with little pride and tended to be dumped rather than placed on the plate. Think of all the prison canteen scenes you have seen in films, where the ladle is brought towards the plate from a great height, followed by the mince landing with sufficient force to spread it across the entire plate, your left hand and the front of your trousers.

Something had to be done.

At a meeting entitled 'stop the plague' held by the OIM, the disgusting situation was forcibly pointed out to the Camp Boss. At first he seemed puzzled, but promised to 'look into it' and 'sort something out'. However, after several days of continuing mince splatter, it became obvious that he had done nothing. Actually this wasn't quite true, an attempt had been made—vests had been washed and were now a tasteful grey colour. It was obvious the Camp Boss was far more interested in selling fags and perfume than in cleaning up the galley.

Enough was enough; we contacted the catering Manager and explained the situation to him, stressing the potential for an outbreak of Bubonic plague. The results were instant and startling; the Camp Boss was sacked and a replacement arrived on the next chopper.

In view of the unusual circumstances, we gathered in the OIM's office to be introduced to him. He came with impeccable references and had been a catering company trouble-shooter in many areas around the world. The only slightly disconcerting thing was that he was clearly gay. I don't want to sound patronising, he wasn't mincing or effeminate, but was, on his own admission, not really interested in the female of the species.

We couldn't disguise our fascination with his immaculately *coiffured* hair; it was wavy, shiny and beautifully cut. Comparing his hair to our own was like comparing Mr Teasy-Weasy to the average men's barber (Short back and sides sir, and something for the weekend?')

The OIM summed it up later by saying with some feeling, 'My wife would kill for hair like that.'

However, our hirsute new manager was as good as his reputation and within a matter of *hours* significant changes began to take place. Clean tee shirts, white trousers and chef type hats all miraculously appeared at the next meal and our man could be seen closely supervising every move in the galley.

It was also interesting to see our pre-conceptions being demolished. The fact that the new manager was gay made absolutely no difference to his ability to light a fire under a bunch of hairy, unkempt galley staff and, without raising his voice, make radical changes to their behaviour. If you're not careful you learn something about social psychology every day.

I remember the Production Supervisor confessing that whenever he was close to our friend, he felt the need to speak in a deeper voice. Silly really, but I guess it reflected our own sheltered upbringing.

Once our trouble-shooter had completed his task, he left the platform (regrettably, not least of all, because he turned out to be a mean snooker player) and a new Camp Boss appeared.

This chap was obviously chosen for his ability to run a 'tight ship' as there were no signs of slipping back into bad habits. He was also a figure of some considerable envy as far as we offshore tigers were concerned. This was because he lived in Majorca and as soon as his one-month tour of duty was finished, he boarded a plane for London and within four hours was basking in sun and sangria in Spain.

To we parochial people, this was the ultimate in sophisticated living and the subject of considerable debate. I recall a number of guys trying to sell the idea to their spouses, but as many of them worked one week on and one week off, it didn't seem to be economically sound. Still, the sums covered many sheets of paper in the control room, as people tried to convince themselves that Majorca was easier to reach than Manchester (which often proved to be true.)

Our new Camp Boss was also credited with the most innovative piece of thinking I have ever witnessed. Not only was his proposal received with untrammelled joy, but he was also credited with hastening the onset of heart, arterial and blood pressure problems among the staff.

It happened like this. Our galley employed a night shift cook whose job it was to bake bread and rolls for the following day shift. At the time, however, we had just lost our permanent cook and our new Camp Boss solved the vacancy problem by bringing with him a real live patisserie chef from a restaurant in Majorca;

as this was the quiet season in Spain, the chef was glad of the money. That first night he proceeded to go into full creative mode.

Imagine our astonishment when, on entering the galley the following morning, we were greeted with tray after tray of pastries. It was unbelievable;

there were chocolate éclairs, cream fruit tarts, Eccles cakes, doughnuts, muffins, buns and biscuits as far as the eye could see. It took some minutes for us to realise that this heavenly spread was all for us. (Note again—no wives, mothers, etc. to rein in our worst excesses.)

Tea break that morning took on an air of reverence as we savoured and sampled to our hearts delight. Actually, 'hearts delight' was an unfortunate simile, as the next morning it was all there again with a different selection, just waiting to be consumed. Could this be true? Once again there were unlimited supplies of sticky cakes for the taking and as we savoured, the questions on everyone's lips were—'What will we do when he goes off duty?' and 'How much will it cost to bring him back?'

And indeed he did come back. All told he made several trips and each time produced a gargantuan number and variety of the most wonderfully unhealthy cakes we had ever seen. Talk about being let loose in a sweetshop. Sadly, the time came when it was time for him to re-open his restaurant and he made his last trip.

Such is the perversity of human nature in the face of the inevitable, that we became rather noble, saying how it was for the best and just as well, for our health's sake.

What pretentious rubbish we spouted, until our cake selection reverted back to a few non-descript scones, of the 'My god, is that a current I see?' variety.

It was interesting to observe just how quickly people became blasé about food. In the early days particularly, there was a superb selection at each sitting, with steak in one form or another always in evidence. Standing in the queue one day with our Production Supervisor, I was taken aback to hear one of the contract welders, on reading the menu board, say, 'Oh bloody hell, not steak again! I'm just about fed up with sodding steak at every meal, why can't we have something different?'

As the mutters of support from other 'gourmets' in the queue died down, my friend turned to me and in a loud voice said, 'Did you hear that bloody fool talk about steak? Well I know where he comes from and I also know that until he came offshore, he had never tasted steak in his life, all his family lived on fish and bloody chips.'

I thought we were in for a slanging match until one of the other welders in the queue agreed, saying, 'Why don't you stop moaning? If you don't want sodding steak, keep your mouth shut and leave the meat to those who are man enough to eat it.'

This had the desired effect, as there was considerable laughter and some surprisingly ribald comments regarding the complainant's manhood, from others in the queue.

During my time on our first platform we took delivery of a large ice-cream machine, supplied as a result of the topic being raised at one of our committee meetings. One of our American friends had lamented the fact that we (read 'he') had very little in the way of comfort food, no popcorn, ice making facilities, coke machines or ice cream. All items which were, apparently, readily available and taken for granted on American installations.

Sid, in a rare attempt to ingratiate himself with our transatlantic friends, managed to procure the machine, which delivered untold quantities of the 'softie' variety of ice cream.

It was located in the dining room, and the ice-cream was available at mealtimes and before the start of the nightly film. Timing is everything.

Gus, the good Toolpusher, was an ice-cream addict and, as he was a senior member of staff, dining room opening hours meant nothing to him. The galley staff were well used to him appearing at regular intervals to avail himself of either a large cornet or a polystyrene cupful and he regularly turned up with one or the other (and sometimes both) at the daily meetings. We used to kid him that what he needed was a machine of his own, a sentiment with which he readily agreed.

This unfortunate remark led to an occasion when a joke we played on Gus backfired. Due no doubt to its popularity and the child-like behaviour of those who think ice-cream is better than custard, the machine broke down irretrievably and there was no option but to ship it ashore and find a replacement.

This desperate turn of events initially went unnoticed by Gus, as the drill crew were pulling a drill string (don't ask, it's a drilling thing) and he had been on the drill floor for many hours.

Anyway, while he was thus engaged, we thought it would be a good wheeze to re-locate the broken machine to Gus's office, in the hope he would be fooled into thinking his dream had come true.

Witness our (childish—I don't think so) delight when he finally returned and on spotting the machine, immediately tried to fill a polystyrene cup. As the awful truth dawned and the cup remained empty, he cursed us in several strange languages, the substance of which meant that he would have his revenge.

We, on the other hand, still thought we had been very clever and the following morning I sent in the 'heavy gang' to remove the machine ready to ship it ashore.

Imagine our surprise when Gus informed them that if they so much as touched the machine, he would personally throw them overboard.

Don't be fooled by the term 'heavy gang', they weren't necessarily heavy individuals; it's just that their specialist skill was the ability to move heavy equipment. Also, as they occupied a fairly low place in the pecking order, they were terrified of Gus—and rightly so.

After having to deal with a mutinous 'heavy gang' I nipped in to see him and asked what was going on. At this, he smiled, a bit like a crocodile, and said, 'You lot think you're clever, well, while you were congratulating yourselves, I've been onto the Drilling Contractor. They say they can obtain spares for the machine and are willing to have it fixed for me at their expense, so sod off.'

Game, set and match to Gus I fear. I still think the stupid idea's was Liam's, but he disagrees.

Another offshore food source was to be found in the sea, but this couldn't be construed as either necessary or immediately consumable. I refer to the quaint hobby known as 'fishing.' As I recall, this new recreation was started by one of our radio operators who very quietly used to fish off the south end of the Platform. What's more, he achieved considerable success and on leaving to go home, could be seen carrying a black plastic bin bag. It appeared to be very heavy and extremely cold containing, as we discovered, about twenty pounds of frozen fish. Apparently, he had an arrangement with one of the cooks, who would allow him to gut and clean his catch and then store the fish in one of the deep freezers. There they lay until it was time to go home.

Sadly, as far as our fisherman was concerned, the secret could not be kept forever, and as soon as word got out, reels started to appear and in the evening men could be seen lining the rails hoping to catch a 'big un.'

What was interesting to us non-fishing onlookers was that each angler soon claimed to have a unique insight with regard to 'the perfect spot' for casting their lines. At first these were closely guarded secrets, particularly when seen to produce a good catch. However is wasn't long before they had an irresistible urge to boast and it was amazing to listen to the garbage they spouted to account for their success, or more usually, lack thereof. When I worked in the pits we used to call these people 'tin pot psychologists' but I really don't know why. Nothing changes.

Some were convinced the heat from the flare boom attracted the fish; others who had no catches that evening were equally convinced that the heat drove the fish away.

Another favourite was the prevailing wind, guys who hitherto had no idea where west was, now became oracles in the movement of fish under the influence of a beneficial westerly wind on the prevailing current.

Then there were the relative merits of nylon line to be argued about, and issues such as its colour, diameter, length of cast and optimum weight. Again, as I understand it, pronouncements were based mostly on hearsay, horoscope readings, lucky medallions and collective opinion, as a substitute for analysis and sound scientific evidence.

To be fair, we should remember that despite all the mystique a fair amount of fish were indeed caught.

Maybe it was simply that the fish thought they were safe from idiots with hooks so far away from land; or maybe it was just that the sea was teeming with fish; whichever, impressive catches were accomplished.

I even have a photo of me posing with a particular fish which one of our contractors caught on the Buoy. It was a Ling, a member of the cod family, and I remember it was all I could do to hold it vertically for the picture. It must have measured at least five feet in length and weighed some forty pounds. (This is not the usual anglers' exaggeration, as I played no part in its landing.) Success such as this did, however, pose something of a problem because there was no way in which the cook would give permission to store it in the freezer.

The fishing craze soon caught on, so to speak, and inevitably fishing clubs were inaugurated. This meant that appropriate club names had to be thought of, a President, Secretary and Treasurer had to be appointed and meetings had to be convened. It was astounding to see just how formal things were becoming, with club rules, fishing spots allocated on a rota basis and catches weighed, measured and recorded. Photographs of significant catches began to appear on notice boards and one or two committed exponents even began to don appropriate fishing apparel prior to venturing forth. This landed gentry look was somewhat diminished by having to wear a bright yellow safety helmet in conjunction with Harris Tweed overalls. The other problem was the difficulty of sticking assorted 'flies' around the rim of a plastic hat.

And then sadly, it all fell apart.

News came in that one or two divers had become tangled in discarded fishing lines which had snagged on the undersea structure and unbeknown to us, were floating about like dangerous seaweed.

Obviously the divers couldn't see the nylon line and they had become caught up in it. Although they weren't in immediate danger, we had to consider the consequences the risk of entanglement could pose.

The result was the Company took the only safe course of action and banned all fishing from our installations. In many ways it was sad but inevitable. It removed what had been a harmless pastime and a change from watching rubbish films. It was a credit to the band of anglers that there were no futile protests; everyone subscribed to the view that divers had a dangerous enough job already.

We never found out what happened to the landed gentry outfits, they just disappeared; although one of the stewards thought he had seen some in our rubbish skip, this was never confirmed.

And so ended another brief but enterprising spell in our never-ending search for harmless amusement.

20

Management Style

I did quite a lot of managing while working offshore; I managed to deal with major problems, I managed to experience helicopter aerobatics, I managed to meet interesting people and I generally managed to get others to do what I asked. I also learned a great deal about 'management style,' which ranged from sheer incompetence to superb leadership. Somebody once said that Managers have to perform many roles in an organisation and how they handle various situations will depend on their style of management. I don't recall who said it, but it sure sounds as though it was a 'Guru' and therefore it must be correct.

Genghis Khan is believed to have remarked, 'My primary managerial talent is downsizing. On my last job I downsized my staff, my organisation and the population of several countries.'

What follows may provide an illustration of the many and varied attempts made to manage a number of genuine but surreal issues. Some turned out to be unintentionally hilarious; others introduced us to so-called 'managers' who had an interesting inability to grasp reality. Many, in their efforts to demonstrate their decision-making prowess, confirmed the comment made by George Bernard Shaw that 'For every complex problem there is a simple solution that is wrong.'

We also learned to work with people of many nationalities and generally there were few real problems—other than the fact that we 'Brits' believed we had the exclusive rights to humour and therefore insults were to be seen as 'tongue in cheek'. At least that's how we rationalised it to ourselves.

Without a doubt, it was Sid's unique management style that gave me the most pleasure. He had an unerring ability to defuse emotive situations and ignore issues which, given time, would simply 'go away'. As this approach was coupled with a sort of world-weary cynicism, I was provided with both entertainment and some fear.

If there was one thing I learned from Sid, it was not to rush in and try to resolve everything at once. He was renowned, and sometimes criticised, for making us provide tangible evidence of a problem. He would demolish the spurious claim that 'We need to do something (anything) immediately,' by forcing us to recognise that just because *we* saw a need to spend £20 million that view may not be shared by management. Often though, what gave me the most pleasure was his way of managing the earth shattering problems we raised at committee meetings.

Management style was very much in evidence regarding the many and varied approaches taken to an event known as the annual appraisal or, to give it its unexpurgated title, 'Performance and Development Review'. This comprised a number of A3 sheets containing exciting headings such as—Appraisal against Targets, Competence Shown in Present Job, Plan of Action and Possible Short-term Development. Having filled this lot in for each employee, you then were required to rank their qualities, estimate their current potential and long-term development possibilities.

This was okay, as long as you bore in mind the need to defend your conclusions with the poor sod, get his agreement to the imaginative screed and send him off happy in the knowledge that promotion was just around the corner. Even if it wasn't.

What we had was a classic example of the difference between aspiration and realism. No matter how many times you did the maths, each recipient was determined to ignore the fact that there were very few supervisory posts available—especially when compared to the number who believed they deserved immediate promotion.

Presenting this stark truth inevitably led to accusations that because it was now December, I hadn't taken into account what they were doing way back in January. This was based on spurious assertions such as that they had worked till midnight on numerous occasions to meet my impossible deadlines, save the planet and increased productivity by 200%. But somehow these modest contributions hadn't been reflected in their annual performance review.

The firm belief that all their efforts, overtime, headaches, heartache, pain and sacrifice had obviously been ignored, was proof that the Company just didn't care. In fact the only thing that seemed to mollify the disgruntled was a promise (which you generally *could* keep) of much more overtime to come in the following year. Thankfully, an appeal to the pocket always seemed to do the trick; such I suppose, is the power of money.

Having your own appraisal was also something of a revelation and one that sticks in my mind occurred when I was to be assessed by an Engineering Manager newly returned from the Far East. He sent for me and waving the appraisal sheets intoned, 'Here, Brian, I can't be doing with this crap, you fill the blasted thing in and then we can discuss what you've said.'

I said, 'Do you mean just my name and other details?'

'No, I mean fill in the whole thing, after all you know better than me what you have been up to all year.'

Actually, it turned out to be more difficult than I thought, due to the fact that if I tried to be reasonably modest so that I could justify what I said about myself, I would be unable to refute the conclusions.

That was when I realised he had been more cunning than I had initially given him credit for, the crafty sod.

I suppose it was just a combination of old age and good luck that I managed to miss out on the new Holy Grail called Psychometric Testing. This miracle system is supposed to assess your ability, aptitude and personality for which scores are given. Its aim is to ensure that you only pick the right person for the job, I keep trying to imagine what might have happened if some Human Resources 'expert' had inflicted this method on Isaac Newton or Winston Churchill. Even more frightening might have been the conclusions drawn from the exercise. Enough ranting, on with the story.

The main contractor on one of our largest 'hook-up' projects was an American company and the two back-to-back Managers couldn't have been more different, both in outlook and management style.

Jerry was from a small town in Kansas and belonged to one of those religious sects so beloved of Middle America. His management style was to seek help from the Lord whenever a problem reared its head, and as problems occurred hourly, we became accustomed to phrases like 'Oh sweet Jesus help and guide us in our tribulations and show us the way forward.'

Now, it wasn't that we were unaccustomed to the name of the Lord being invoked, although I have to admit it was generally in a somewhat different context. You only had to pass the welding shop or the snooker table to hear his name being mentioned, often coupled with blame, for the player's failure to achieve a finish on the black.

Jerry wasn't the Elmer Gantry, fire and brimstone type of religious nut; in fact, we would probably have preferred this to his rather sanctimonious approach. As Liam remarked, after Jerry had just made a particularly lengthy appeal for the Almighty to fix a damaged turbine impeller, 'If this works, I'm going to ask him if he can fix my bloody up and over garage door.'

Jerry's opposite number however, had a different approach both to life and man-management. He was a legend in his own lifetime who introduced himself modestly as 'Diamond Jim'. He had worked for the company for many years and prior to this latest appointment, had been a Barge Master working in the Bayou swamplands in Louisiana.

His management style was of the John Wayne variety and seemed to owe much to his having spent too long in the close proximity of alligators. To say that his approach was a surprise to trades union officials from our northern climes would be something of an understatement.

He wore pointy toed, Cuban-heeled cowboy boots, deeply embossed with prancing horses. The toes had silver ends and there were leather thongs hanging from the rims. A pair of skintight Levi jeans, a thick check shirt and, horror of horrors, an aluminium safety helmet, completed the outfit. We couldn't wait to see the OIM's reaction.

All efforts to make him wear safety boots and a plastic helmet were met with fierce resistance. Jim's responses to all safety-based appeals were delivered in a largely incomprehensible southern drawl. The gist of it was that he thought we weren't 'big oil and gas men' and consequently he referred to us as, 'ya'll a bunch of Goddamn sissies.' His tirade would be liberally interspersed with references to someone called 'Jesus. H. Christ', but in a markedly different manner to that used by Jerry.

We were agog for information; who was this gunslinger? What would our Newcastle welders make of his style? Why was he called 'Diamond Jim'? We were soon to find out.

Now, as I said earlier, our experience with jewellery was largely confined to a Seiko watch, the odd (very odd) daringly worn earring, usually a ladies 'sleeper' or a brass curtain ring, and to complete the ensemble, a St Christopher medallion on a nine carat gold chain obtained from H. Samuel (jeweller to the masses.)

Jim's idea of personal embellishment bore no resemblance whatsoever to our attempts at being manly objects of desire. On any average day, Jim would be wearing enough diamonds to give De Beers cause for concern. His ensemble would typically consist of a huge gold ring with a flat top upon which were his initials 'DJ' set out in diamonds (as I said, he actually referred to himself in the third person as 'Diamond Jim'.) On his wrist was a genuine Rolex watch with each numeral tastefully highlighted in diamonds and to provide balance and cohesion, a ring of diamonds mounted around the bezel. The strap was similarly embellished with the usual 'DJ' picked out on the clasp.

His jeans were held up with a substantial tooled leather belt with, yes you've guessed it; the buckle liberally adorned with yet more diamonds. The man was a veritable walking jewellers shop.

Oh, and in answer to what we thought was a sarcastic enquiry, he confirmed that the metal tips on the toes of his boots were silver—obviously!

Jim was not the least bit reticent regarding his finery and was more than happy to show it off given half a chance. He said it represented a lifetime of working in the oil industry and assured us that all the diamonds were genuine.

I remember Bert, in an attempt to engage Jim in conversation, asking whether he had any other adornments that he had left behind.

'Goddamn right, I have,' said Jim, and proceeded to describe cufflinks, numerous belt buckles and pairs of sunglasses with his initials picked out on each side. But someone had told him that there was no sunshine in the North Sea and so, to his regret, he had come without them. I looked meaningfully at Bert, hoping he would change the subject, as there was only so much we poor relations could take of this unassuming American swine.

I remember we were chatting one night and the subject once again turned to Diamond Jim. This time we spent a happy hour trying to work out just how much Jim's accoutrements were worth. Mostly, we only got as far as estimating the cost of the Rolex; the value of the diamonds was just a shot in the dark. However, what we were all agreed on was that the entire collection would have bought us all a new house and a Ford Cortina (the car of choice for the discerning family man.)

It was Liam who really summed up our feelings, when he said that Jim was the only man for whom it would be worth bringing in our diving vessel to retrieve, should he fall overboard.

We all thought this sentiment to be exactly right, especially when you consider the going rate for the vessel at the time was a snip, at about $25,000 per day.

As you might have guessed by now, Jim had a supreme ego and although he could be considered a 'Prima Donna' he wasn't arrogant or aloof. This, when you weren't a victim of his entrenched opinions regarding all things American, made for a likable and interesting person to work with.

One excellent illustration of the esteem in which he believed himself to be held, occurred one day as I left a progress meeting in the main contractor's office. Going down the corridor, I bumped into Susie, the Project Manager's Secretary, and stopping to chat, I mentioned that we were slowly getting used to working with Diamond Jim.

At the mention of his name, she gave a slight shudder and said, 'Huh, do you know what the idiot bought the girls in the office for Christmas?'

As I didn't care to speculate, I said, 'No idea, Sue, what was it?'

She stared at me with one of those withering looks which women sometimes reserve for men who fail to understand and said, 'The conceited swine gave every girl a signed photo of himself in a silver plated frame.'

I have to say I was somewhat taken aback. 'Why would he do that?'

'He said it would look nice on our desks and would give us all something to remember him by.'

I thought bloody hell, ask no more questions, and saying how nice it was to see her again, I shuffled off. Somewhat illogically, I couldn't help thinking how association with our modest present giver might further diminish me.

I mentioned this conversation with Sue to the lads on my return offshore and Bert summed up our feelings by commenting, 'Nothing bloody well changes, it was the same in the Second World War when our chaps used to lament that the Yanks were 'over paid, over sexed and over here.'

There were nods of agreement all round, (even though we were all at primary school during the war) and we went off for a well-deserved dinner.

We had a number of interesting experiences with budding OIM's during our early days and I remember one in particular, an ex Royal Naval Officer, whose approach to man management illustrated the massive gulf between his past life and his new career.

Each afternoon at three o'clock it was our tradition to repair to the OIM's office for afternoon tea and sticky cakes and on one particular occasion, our tyro OIM arrived late and trying the empty teapot, said, 'Oh I say, has all the tea gorn?'

We were happy to confirm this was the case and, seeing his evident distress, suggested he nipped off to the Galley for some more.

Somewhat disconcerted at this new self-help regime, but willing to try, he went into the corridor where, spotting a steward mopping the floor, he decided to issue an order.

'I say, you there, bring me a fresh cup of tea, there's a good chap.'

Unfortunately, this didn't quite have the desired effect and he was subjected to a tirade, which implied that fetching tea was not terribly high on the steward's list of priorities.

'What? Are you talking to me? Don't you think I've enough to bloody well do? You idle sod, get your own rotten tea! What do you think I am, your bloody servant or something?'

As the steward warmed to the task, he began to punctuate his increasingly more disgusting remarks with positive action. The immediate result was that the mop was flung in a lazy arc down the corridor, whilst at the same time spraying mucky water in all directions. This was followed by an almighty crash as the bucket, now being used to provide suitable emphasis to another tirade, was kicked across the lino into the far wall.

Fortunately, by this time, we had arrived on the scene and stepping gingerly into a pool of dirty water; some of us tried to pacify the steward, while I dragged Captain Bligh back into the office.

It was obvious from his demeanour that he believed he had just witnessed a mutiny, but was lost as to the reason why. 'What did I say? Is he mad?'

I tried to explain. 'Yes, you could say he's mad but unfortunately, he's mad at *you* and not certifiable. What he was trying to tell you in his own inimitable style, was that he had a job to do which certainly didn't include waiting on your good self.'

Looking like a man who has just entered a completely alien world the crestfallen Bligh said 'I don't think I'll bother with tea today.'

On another occasion, I was chatting to the Radio Operator when one of our Construction Engineers went past. Just then our trainee leader spotted him and coming over all leader-like said, 'I say, are you one of the construction hands? I need to get someone to clean up the welding shop, be a good chap and see to it will you?'

There was a sort of stunned silence, the Radio Operator tried to slide quietly into his room, the Engineer stopped dead and, just before he went ballistic, I dragged him into my office.

'Who the hell was he talking to and what's all this about me being a *hand?* I'll kill the daft sod!'

Again I tried to explain. I was in danger of becoming a sort of offshore agony aunt.

Later on, I said to our walking disaster area, 'Look, if you don't want to find yourself tipped into the sea by person or persons unknown, you've got to stop referring to people as 'hands'. The guy you issued your clean-up instruction to happens to have a First Class Honours Degree in Engineering from Brunel University. Just because we all wear jeans and frayed polar jackets doesn't mean we have no status.'

Sadly, I could see that this radical departure from all he held dear in officer–other ranks relationships was a total mystery to him. Clearly we were into a protracted learning curve—if he lasted that long.

In the late eighties we were 'blessed' with an increasing number of Senior Managers who, having been chosen as having 'fast track' potential (to the cynical, this translated as a triumph of useless IQ over ability,) were sent off to the United states, where they were exposed to what were considered to be the world's foremost (and expensive) Management Consultancies.

These consultancies employed teaching methods reminiscent of the Great Revolution in China, the main difference being that the Lecturers wore Brooks Brothers' suits instead of Chairman Mao smocks. An escapee manager told me that a major brainwashing technique employed by one lecturer was the chanting of a mantra that went 'Are you guys with me, okay, let's really challenge those paradigms.'

Now, I have to admit to a grammatical weakness here, because no matter how many times I hear the word 'paradigm', or see it written down, I still have no idea what the hell it means, or when I should use it without sounding like a prat. This is despite repeatedly looking it up in the dictionary and trying desperately to memorise the explanation. Sadly it is to no avail, ten minutes later I can't remember a damn thing and realise once again that I will never be able to join that exclusive 'How to slip the word paradigm into a conversation' club.

I bet Jeremy Paxman doesn't have that problem.

I guess the word paradigm suffers from a similar problem to that of algebra, in that millions of people were subjected to it in school, but have managed to change light bulbs and raise whole families without ever having to use it.

What we hadn't realised, whilst our bosses were being indoctrinated in the American way of business, was that we were witnessing the birth of 'management speak'.

At the time, we were all too familiar with the tendency for managers to indulge in 'euphemism speak' and many happy hours were spent trying to interpret their coded messages. You know the type of thing; politicians use it all the time to avoid actually answering a question.

> That's very interesting, means—I disagree.
> I don't disagree, means—I disagree.
> I don't totally disagree with you, means—You may be right, but I don't care.
> You have to show some flexibility, means—You have to do it whether you want to or not.
> We have an opportunity, means—You have a problem.
> You obviously put a lot of work into this, means—This is awful.
> Help me to understand, means—I don't know what you're talking about, and I don't think you do either.
> We need to syndicate this decision, means—We need to spread the blame if it backfires.

You can see how crucial it was to be able to interpret the actual message, whilst at the same time, remembering to use the project manager's golden rule—if a subordinate asks a pertinent question, look at him as though he has lost his senses.

And people try to convince you that cryptic crosswords are stimulating.

There is one classic example that still sticks in my mind and this time the proper translation came from another manager with whom I was attending a Senior Management briefing.

Part way through the presentation the chap in charge said, 'So gentlemen, the situation leads me to believe we have to put on our marketing hats and leverage our resources.'

Although, obviously, we all nodded in agreement, there were many surreptitious looks of consternation and bafflement, until my friend whispered, 'What the bastard really means is that we have to put ethics to one side and screw our contractors before they screw us.'

What we hadn't realised was that this new era of 'management speak' represented a quantum leap in gobbledygook and would leave us nostalgic for the simple days of good old punitive management, even when liberally laced with euphemisms.

What was coming would raise bafflement to undreamed of heights and create a whole new science, whereby managers would say words which, when taken individually, seemed familiar, but when strung together made interpretation impossible.

Words began to appear in documents, like our friends 'paradigm' and 'solutionise'

As if the new words weren't bad enough, they were then strung together to form equally meaningless phrases. Some have gone down in history, others refuse to die and heaven help us, are still trotted out whenever exhortation is called for. Again, we have some classics, which to the uninitiated, must seem to have little to do with getting oil and gas out of the ground. Just imagine a bunch of hairy engineers suddenly being confronted by phrases such as:

'Let's run a few ideas up the flagpole and see who salutes...'

'We are going to be raising the bar'

'We need to ensure we're all singing from the same hymn sheet before we push forward with this'

'Let's cut to the bike-sheds'

'We need to think out of the box'

'On a going-forward basis.'

However, my all time winner of the gobbledygook award must go to a wonderful example, which has stayed in my mind for some twenty odd years. A visiting consultant said it on a Project Management course I attended. At one point, he asked us to consider the need for innovation when scheduling sub-contractors.

'No need to re-invent the wheel going forward. It's not rocket science; all we need is a stand-up meeting so we can start the hard yards on some blue-sky thinking. I'll touch base later this pm at our catch-up session'.

I think what he was trying to say was, 'I'll meet you this afternoon to see how we might approach the problem in a new way.' At least that's what I thought he meant, but to this day I'm still not sure.

There are some things that are just too precious to forget.

What I enjoyed far more than the 'double-speak' employed by the so-called 'McKinsey men,' was to be involved in the age-old battle between managers of radically differing styles. This could be entertaining, but when you worked for three years with two guys who had diametrically opposed approaches to management, life became somewhat more interesting than any one person had a right to.

I speak from experience, as it was just such a situation in which I found myself on one big Project, when in charge of Commissioning. What made it even more 'interesting' was that I worked for both of them at the same time.

The Project Managers of whom I speak worked 'back to back', two weeks on and two weeks off and someone, who didn't have to make the system work, came up with the brilliant idea that I should work my first week of duty with Harry and my second with Vince. The logic was that I would provide continuity and smooth over any potential handover problems. This seemed reasonable until I realised just how different their management styles were. Consequently I became a sort of acting unpaid negotiator, with a task comparable to single-handedly trying to get both North and South Korea to love America.

Its not that they were incompetent, far from it, they both had sterling qualities; the problem was that Vince was from the old punitive school of management, honed to perfection over many years in foreign parts. This was where, should you inadvertently drown a number of workers, you simply sent for another lorry load. Or so the saying went and Vince certainly did nothing to dispel the illusion.

Whilst Harry was of the new, highly organised, forward planning school of management.

Fortunately, they both had one goal in common, to get the platform built and handed over and, as I often appeared to frustrate this goal, they also had one common enemy—me.

I mentioned earlier that Commissioning, which I was responsible for, was looked on with a good deal of suspicion by the construction group, and as far as both Harry and Vince were concerned, we were deliberately obstructing their efforts to finish the job and sail off to pastures new.

Their basic argument was that once they had built something, it was *finished*. But frustratingly for them, my team seemed determined to thwart this erroneous view by making outrageous demands to see the plant operate consistently before declaring it fit for handover to Operations.

The scene was, therefore, set for three years of endless fun, paranoia, derangement and a growing ability to adopt Machiavellian tactics on my part.

A typical tour of duty for me would run something like this; I arrived at the start of Harry's second week and would be briefed on progress so far. This would be illustrated by referring to a complex Critical Path Analysis Chart, which covered an entire wall in Harry's office. The chart was sub divided into strategic plans, tactical plans and latest schedules with lots of coloured boxes and arrows, all of which was created by Harry's Planner, a quiet guy named George. After a detailed briefing, I would settle down, see the lads and work happily away for the next seven days.

Then, a week later, Harry would hand over to Vince and we were now under a totally different managerial regime.

Vince worked to a timetable that was uniquely his own and had nothing to do with our requirements. On arrival, he would generally repair to his cabin and have a lengthy nap. This arrangement was preceded by a demand for the Cabin Steward to 'Get rid of that new fangled duvet' on his bed and replace it with a white sheet and two blankets.

The first time Vince did this he caused a furore because the steward he had so instructed, being nonplussed by the demand, reported the matter to the Camp Boss, who came over all macho and flatly refused. He argued that he had fought long and hard to obtain nice duvets (with matching pillowcases,) how pleased everyone else was to have such improved sleeping arrangements and how his laundry bills had been slashed.

The steward nodded in agreement, but on taking a sidelong glance at Vince, realised this wasn't going to be as easy a victory as he had first thought; primarily because at this point, Vince, having so far remained silent, growled at the Camp Boss, 'Right, you bastard, listen carefully: either I get my sheets and blankets before tonight or you're sacked.'

This new approach to the process of negotiation and amicable settlement took the Camp Boss somewhat by surprise, but on recovering, said the wrong thing.

'We'll bloody see about that,' he snapped and, making a rare appearance outside his lair, nipped in to see Walter, who was our OIM.

Walter, the Camp Boss and Vince then entered into a series of delicate negotiations based on bribes, threats and offers of resignation, the upshot of which was that Vince rescinded the sacking promise and the Camp Boss rang around the other installations to secure sheets and blankets and blamed the steward for getting him into this mess; then Vince went to sleep.

After his nap, thoroughly rested, Vince would appear in the office at about four o'clock and summon us to his presence.

Invariably, his first edict was to was to wave at Harry's detailed wall charts and say, 'Right George, get that crap off my wall.'

I remember the first time, George presumed he hadn't heard correctly and made no move to comply; probably because he and Harry had spent many hours refining, colouring, copying and distributing copies of the good words to everyone.

Vince, however, using his direct management approach, remedied George's misapprehension by leaping from his desk and ripping the whole lot off the wall.

I thought George was going to cry.

We all (cautiously) remonstrated with Vince, saying the ruined flow chart represented the most up to date picture of the Project. Vince just snorted, 'Stuff that, what I want are status reports from you lot and then I'll tell you what needs doing.' I guess this was what you might call the 'I'm in charge' method of managing complex projects.

Obviously after a short time we devised tactics to alleviate George's growing paranoia and just as Harry left and before Vince arrived, we rolled the charts up and stuffed them in a cupboard, to be retrieved and re-hung just prior to Harry's return.

One dread we all had was that Harry would find out just how little notice was paid to his life's work when he was on leave. I thought the game was up when one day he mused, 'It's funny that Vince never updates the charts during his shift.'

I mumbled, 'Vince thinks it better and less confusing if updating is left to one team.' I was beginning to learn.

Vince also worked a totally different shift pattern to the rest of us. Whereas we, foolishly, were on the job at seven am every morning, he would rise at about ten, meander in to the galley, ignore the fact that it was closed, cadge a cup of tea and a cake, and then take a leisurely stroll to his office.

He would then set about catching up with events. This required him to shout a lot at anyone in the vicinity, write vitriolic notes to 'those clowns onshore', countermand some key edicts from the Project Director, and then demand an immediate meeting with the Main Contractor.

All this managerial behaviour brought us to lunch time, after which Vince would slide quietly off to his cabin to 'read some important papers', during which time he was not to be disturbed.

Other than feeling somewhat jealous, we didn't mind too much, as it meant we could get on with our work undisturbed until Vince resurfaced just in time for Walter's three o'clock meeting, where tea and a nice selection of cakes would be on offer.

Initially we made the cardinal error of passing the cake tray to Vince so that he could have first choice; this was like expecting a crocodile to politely nibble on your arm. Another lesson learned on the advantages of ignoring seniority.

The downside to Vince's unique work pattern was that just as *we* were finishing our normal twelve-hour shift, *he* was at the peak of his powers. The thing we dreaded most was a summons for all Department Heads to attend one of his notorious 'update' meetings. Nothing wrong with that, except they invariably started at ten pm.

It wasn't unusual for there to be about twelve people at these meetings, all looking shattered, with Vince blissfully unaware (and caring even less) that we were not exactly pleased to be there. As we gathered by the coffee pot, there were whispered mutterings that 'Somebody should tell the silly old fool that we're knackered and he should hold his meetings at a civilised time.'

The whisperers were also heard to utter brave phrases such as 'you tell him', it's not my job', are you mad' and so on.

The most heartfelt comment regarding this situation was made by the Ventilation Engineer, who said 'I've worked for the daft sod for three years and in all that time I have never seen a film right through or ever finished a game of snooker.'

It was at one of these meetings that I inadvertently aroused Vince to fury. The room was packed with all the construction electrical, mechanical, piping, instruments and ventilation engineers, with me as the only Commissioning representative.

Vince was once again trying to find out how near to completion we were and his method for obtaining this information was to jab a finger at each Engineer in turn and demand, 'Right, where are we with you lot?'

Now, as most of the group where tired and certainly had no intention of passing on bad news, each nominee would provide Vince with a wildly optimistic status report. I listened with growing dread as the overall message became clear. As far as they were concerned, all construction and installation work had been finished on time and with no discernable problems. This meant of course, that the only group who could now be accused of holding up the entire project was the Commissioning Department.

Slowly but inexorably, the jabbing finger worked its way towards me and, as the rest of the members relaxed or nodded off, Vince said to me, 'Okay, that's just great, now let's hear the good news from you.' You could see by his clever use of the words 'good news' that he certainly wasn't expecting me to ruin his party.

To this day, I don't know what made me say it, but realising I had to dispel the view that all was well and explain that what he had been told so far was, in part, wildly optimistic and in general, a pack of lies, I said, 'With respect Vince, what you have just been told is a load of crap and we're far from ready to start commissioning'.

I was just about to start explaining why, when Vince went red, then purple, then banged both arms down on his desk and shouted, 'Don't you bloody well 'Due respect' me, you bastard, I know what you are up to—you think if you use lawyer's jargon you can insult me and get away with it!'

To say this tirade put me off my stride would be something of an understatement. Here I was full of adrenalin in readiness to defend my contention that everyone in the room was lying except me, only to be met with what I took to be a distinct lack of gratitude on Vince's part.

Suddenly, as far as the majority of those assembled were concerned, the meeting had become very interesting and this better than average outburst from Vince cheered everyone up—everyone that is, except me.

Sadly, in my haste to correct Vince's assertion, I then made things worse by saying, 'I thought 'with respect' meant that the recipient was being regarded with esteem.' In hindsight, this pathetic attempt to retrieve the situation, turned out to be a disaster of the highest order.

However, just as Vince's eyes started to bulge again, I was saved by the Contractor's Engineering Manager shouting, 'Great stuff! Keep going Brian; we're all off the hook and you're digging a bloody big hole for yourself!'

Support of that kind is not to be found every day.

Fortunately, this timely intervention relieved the growing tension and Vince calmed down, although he continued to glower at me.

I have to say that after all this throwing of toys out of pram, Vince waved his arms around the assembled group and said, 'I thought the reports from you bastards were too bloody good to be true. Now Brian, what were you saying about this lot talking crap?'

Time to extract my revenge, which had to be done carefully as I still had to work with the sods.

Remember Diamond Jim? He came over all macho one day and made one of the most potentially devastating management decisions we ever encountered and for which I was the unwitting catalyst. It was also a perfect illustration of the saying 'Two nations separated by a common language.'

We were about to commission a major section of the oil export system and were gearing up to work through the night in order to run the main pumps at full speed for twenty-four hours.

Obviously, there were a number of last minute hitches, one of which was that a section of high-pressure pipe still had to welded in before we could make a start. This involved the use of highly qualified welders, who could easily be identified by the fact that they all wore caps made out of cotton with polka-dots all over them. Not, you will admit, the most convincing picture of hairy manhood. The welders were accompanied by a support team of pipe fitters, welding rod warmers, heating blanket handlers and riggers. And you thought welding was easy!

Anyway, we were waiting for them to complete the last section and as my lads were becoming restless, I paid a visit to the site. It was about three o'clock in the afternoon and to my surprise there was no sign of anyone 'striking an arc' (more technical terms for you.) Carrying on with my search for big hairy chaps with extra long arms and funny caps, I came to the 'Smoko'.

I know this will come as something of a shock, but a number of people on board smoked cigarettes. The management (us) had to control when and where this habit could be indulged without the risk of blowing the whole installation out of the water.

As we weren't in production at the time, we had placed a huge portacabin on the deck and rigged it up with a self-closing door, forced ventilation, lights and tea urns. This was universally known as the 'Smoko', and people with any regard at all for their lungs kept well clear of it. However, in my search for the missing wild bunch, I passed the door just as it opened and saw, through the haze, the polka dot hats and their cohorts all enjoying a brew, a fag and a cough.

The scene reminded me of my early days when I travelled to work by tram and the worst thing that could happen was to find there was no room inside and you had to go upstairs to where the smokers were situated. There were two major disadvantages to this, one was that you couldn't see an empty seat because of the thick smog and the other was that you couldn't hold a conversation because of the noise made by the smokers as they lit up, inhaled and coughed. We are not talking about amateur coughing here; this was loud, racking and could last for minutes on end. The other thing preventing conversation on the part of the smoking fraternity was the need to remove accumulated phlegm as soon as they got off the tram. My mate at the time coined a wonderfully descriptive phrase that summed up the spectacle perfectly.

He said he watched in amazement (and some alarm) one morning when one bloke, on alighting from the tram, proceeded to cough and spit the like of which had not been seen before. I remember to this day his brilliant description of the chap's attempt to remove the accumulated phlegm.

'His cough was so extreme, that bent double and heaving and straining, he brought it from the back of his eyes; it was bright yellow and laced with blood and bits of lung.'

Enough of the social history, back to the plot.

Now, as the welding team didn't work for me directly, protocol required that I bring any complaints to the attention of the Contractors' Manager. Not wishing to cause any more ill feeling than necessary, I decided to consult Walter.

We were in his office debating the best way forward when Diamond Jim came in for a chat. Just the job, I thought and proceeded to express my disappointment regarding lack of progress and noted that that the Smoko was full of bodies, although it was outside of the designated smoking hours.

'Goddamn it!' exclaimed Jim, 'Leave this to me.' And off he went. To our surprise he returned very quickly and I thanked him for his support, thinking the welders were now all back at work. He then dropped what might be called a 'bombshell': 'No problem, it was soon sorted out, I've fired all their lazy asses out of here.'

Walter and I looked at each other, both staggered. I ventured, 'Jim, I guess what you mean is you've given them a good bollocking and they are all back on my pipe job?'

'Sure have,' chuckled Jim, 'But they don't work here any more, the lazy limey bastards are all fired.' With that he smiled the smile of a man who has solved all our problems and sat down to eat a cake.

Walter was the first to recover, 'Jim, please tell me you're joking, you haven't really sacked them have you?'

Jim looked a bit puzzled, having, as he thought, explained his industrial relations strategy in precise terms.

'I told you,' he said, 'It's a done deal; they're off on the next chopper.

Why, what's the problem?'

'The problem,' said Walter, slowly, 'Is that you just can't just *sack* people in the UK. You've got to go through the correct employee-relations process. You have to give written warnings, final warnings and formal notice of reasons for termination. If not, we could be accused of wrongful and unfair dismissal and could end up being taken to an industrial tribunal. *That's* the problem.'

I was impressed; obviously the recent course on employment contracts had not been wasted on Walter. I particularly liked his use of the words 'wrongful' and 'unfair'. While he gathered himself for the next lesson and no doubt mentally trying to word the telex to base, Jim stared at us in total bafflement.

'Goddamn it, you guys asked me to sort it out and I have and now you're giving me grief just because I don't pussyfoot around with these bastards. What the hell do you want?'

Walter was just about to answer at great length, when the Radio Operator put his head around the door and said, 'You might want to listen in to these inter-platform phone calls.'

I followed him out and as he turned up his radio receiver, my worst fears were realised; the contractor's Union Representative on our platform was informing his opposite numbers of the unwarranted calumny visited on his members and suggesting that the brothers on all the other installations might want to consider showing solidarity with them. You don't forget militant jargon like that in a hurry.

It was obvious, at least to me, that we were on the verge of a major cock-up, which would be received with little enthusiasm and lots of blame by the Managers on our sister platforms. Telling the radio operator to keep me posted, I nipped back into Walter's office to convey the good news, just as he was trying to convince Jim that he should re-instate the Tolpuddle Martyrs right away.

Jim went apoplectic at the thought of backing down and insisted on the need, as he still saw it, 'to teach the lazy bastards a lesson'.

I said, 'Jim, don't you bloody understand? You're just about to bring the North Sea to a standstill. You'll go down in history as the only American to do more damage in the world than President Nixon.' For a long moment we stared at each other, both at a loss and with the realisation that my courses on negotiation skills had singularly failed to include the possibility that we might one day meet up with a kamikaze American.

After some heated debate, we came up with a face saving brainwave. We would ask Allan, Jim's Team Leader to convince the troops that it was all an unfortunate mistake, get him to rescind the decision but slip in some suitable warnings and the threat of dire (but unspecified) consequences should abuse of the Smoko occur again.

In other words he would have to carry out a major bluffing campaign.

Walter put out a call for Allan to come to the office, while in the meantime I tried to convince Jim it was the only course of action open to us if we were to come out of this alive.

Fortunately at that moment Allan arrived and, blissfully ignorant of the unfolding drama, sat down and picked up a cake. Walter explained what had happened, how Jim had applied the ultimate sanction to the workforce and how we were in danger of a major industrial conflict.

Allan stared at his boss in horror but still unaware of the crucial role he was about to play, said, with masterful understatement, 'I thought it was pretty quiet when I came through the modules.'

Walter then explained the game plan, which, he said, simply required Allan to meet the sacked workers and convince them that they had failed to grasp the underlying meaning of Jim's message.

He should say there had obviously been a failure on their part to see that Jim had been exaggerating when he said they were all sacked. This had just been Jim's way of bringing some good old American humour into what was after all, just a 'telling off' (or 'bollocking' as it is sometimes referred to.) Unfortunately, they had misunderstood and had over reacted.

Simple really.

Allan, showing a growing sense of alarm regarding his forthcoming suicidal role, bit deeply into the cake. Possibly due to the realisation that Walter and Jim had just shafted him, he unfortunately forgot to remove the greaseproof paper. The immediate effect was that he choked, then coughed; we were all covered in bits of cake and paper and Jim, realising he might be off the hook, offered Allan a cup of tea.

On my arrival on the platform a couple of trips later, Walter introduced me to Harvey, the new Contractor Manager. He was very tall, spoke with a soft Bostonian accent, was well dressed (in a suit, would you believe) and was extremely polite. He told me how he looked forward to working with us and that we shouldn't hesitate to call on him day or night should the need arise. Suddenly it dawned on me that this might be the end of an era, so I said 'Where's Diamond Jim?'

There was along silence and then Harvey said, 'Ah, see now, we have extensively reviewed our organisational framework with a view to equalising our employment opportunities and we felt that Jim would more ably play a key role in our new interpersonal communication strategy by moving him back to the barges in Louisiana.' And with that he smiled.

Talk about euphemisms? What he really meant was that poor old Jim had embarrassed his Company once too often and had been put out of harm's way.

Despite his unique management style and supreme ego, I knew we would all miss him. We were witnessing the arrival of the new breed of 'faceless' Managers.

21

Nothing Ever Happens

There is a saying, attributed to the Chinese and said to be a curse which states, 'May you live in interesting times.' I understand the Chinese don't recognise the saying, but allow that it may be a translation of 'It is better to be a dog in peaceful times than be a man in chaotic period.'

As I don't understand a word of their version, I generally contend that the first one is real; anyway, when prefaced with 'Confucius says', it sounds positively transcendental (whatever that means.)

Overall, my time offshore was interesting in the positive sense; I enjoyed coming to work, I enjoyed the company of like-minded chaps and I enjoyed lots of custard. However, there were a couple of occasions when the Chinese curse became accurate to the degree that you wished they had stuck to being inscrutable and had concentrated on building walls.

Once we were a couple of years into commissioning one of the biggest and most complex of our concrete platforms and had so many engineers and contractors on board that we had a semi-submersible drilling rig moored alongside for use as offices and sleeping quarters. Access to and from the platform was via a bridge connecting the two facilities. You can imagine the logistical nightmare keeping track of everyone's whereabouts, arrival and departure, cabin allocation, feeding locations and lifeboat stations. This enviable task initially fell to the Medic who attacked the job with a singular lack of enthusiasm, resulting in lost personnel, infighting, accusations and threats to the body. We soon realised that what we needed was a dedicated, competent, miracle working Administration Officer.

Enter Admin Al.

Al was ex Royal Navy; he had been the Bosun on one of Her Majesty's aircraft carriers. I understand this was one of the most important jobs on any ship, as the Bosun was responsible for discipline, piping the bigwigs on board and ensuring that every man was correctly on watch and at his station. (The only regret I heard Al mention was that although he had the power to carry out floggings with the cat-o-nine tails, the Royal Navy had banned the practice).

Al's recruitment was an unbelievable stroke of luck. He slipped into our admin job as though born to it, he would countenance no arguments from anyone, he was singularly unimpressed by rank and never, ever, raised his voice.

According to Al, compared to his job on an aircraft carrier, this was a doddle.

Al looked harmless, being of average height, average build with an average voice. However, his ability to instantly quell gripes and arguments about cabin allocations and flight times was astounding. I don't know how he did it, but I bet things would have turned out differently for the motor industry if Al had been in control of British Leyland.

As well as administration, Al was also in charge of a small team known as the Helideck Crew, whose job it was to prepare the deck for landing and take-off, re-fuel the helicopters, guide passengers to and from the helideck, feed the Pilots and ensure that fire-fighting equipment was in operation.

Why am I telling you all this? You'll see the relevance in a minute, but before that, let me fill in some background details.

The first things to be commissioned on a platform came under the heading of 'Life Support' and, as I mentioned earlier, we had to have a certificate to verify that all safety related issues were resolved, before we could bring a full crew on board.

The most important life support system (after the Deep Fat Fryer) was the Fire Fighting System. This consisted of a ring of pipes running throughout the platform fed by a huge fire pump located in one of the concrete legs. One of the distribution pipes ran up the inside of the accommodation on to the helideck, then it branched out and ran across the top of the living quarters to feed the sprinkler heads in the living quarters and offices.

It just so happened that while we were commissioning a Service water pump, a problem was discovered with the water intake, which was literally a hole in the concrete leg below the water line to which the inlet of the pump was connected.

In order to see what was wrong, we arranged for a diving vessel to come alongside so that a diver could go down and see if there was a blockage or damage to the grille covering the intake. Due to the volume of activity taking place during the day we decided it would be safer for the diver to carry out his survey at about ten pm.

Having checked our communication systems, Jim (our Mechanical Engineer) and I descended the leg to the pump site. Reports on the condition of the inlet were relayed from the diver to his Superintendent, who then passed the information to Jim, so we could decide what action to take and hopefully, sort out the problem.

And this is where, as they say, the story really starts.

While Jim and I were happily engaged in the bowels of the platform, chatting to the diving team, a long, long way above us, something had just gone horribly wrong. Unbeknown to us, the eight-inch diameter fire main parted at a joint immediately above Admin Al's cabin. This was a significant event in itself, but what turned it into a disaster was the fact that we had fitted a low-pressure sensor to the pump controls. This meant that if the unit sensed a drop in pressure in the pipework, the fire pump would start automatically. Genius or what?

So, when the pipe burst and the pressure dropped, Al and his team were subjected to the full blast of an eight inch pipe shooting freezing cold sea water into their room at a pressure of about eight bar, or 120 pounds per square inch. Have you *any* idea just how much water that is? No, well let me assure you it is more water in a millisecond than you ever dreamed possible, it is relentless, it makes a noise like the Victoria Falls on a bad hair day and is a credit to the skill of the fire pump manufacturer.

It gets better (a relative term in the circumstances.) An Operator whose job to was to take readings, fill out the logbook, issue permits, doze a little and read newspapers manned the Control Room that evening. On this particular occasion the guy on duty, being in alert mode, noticed that a red light had appeared on his control panel indicating that the fire pump was running.

He then made two equally false assumptions; first of all that the helideck crew were testing the fire hydrants; and then secondly, that if it wasn't them, it must be Jim and I running a test. After this exercise in arriving at completely wrong conclusions, he decided all was well and went back to an article discussing the relative merits of the Boomtown Rats versus Olivia Newton John. We gleaned this choice information during the subsequent investigation, while he was locked away for his own safety.

In the meantime, comfortable in the knowledge that this was what it was born to do, the fire pump was now sending massive amounts of water into the accommodation.

We need to pause for a minute while we begin to comprehend the scale of the calamity.

It was now about eleven o'clock at night; most of the day shift had turned out the lights and gone to sleep. Jim and I were still down the leg thanking the divers and whistling cheerfully; the night shift were about to have their dinner in the canteen and the Baker was just putting freshly made bread and rolls into the oven.

Suddenly, there was an almighty crack as the pipe parted and Al's cabin began to fill with very cold seawater, much to the dismay of the occupants.

In the dark and in their underpants, they either fell or were washed out of their bunks and in some disarray, managed to force open the cabin door, thus allowing the water to flow into the hallway, along the corridor, into the other cabins and down the central staircase like a waterfall.

Remember, the fire pump was still pushing out water at a frightening pressure and quantity. In quick succession, people were being soaked, cabins were filling, ceiling tiles were turning to mush, wall sockets were short-circuiting and the lights were going out.

Al said later, that on being initially deluged, he thought 'Bloody hell, the ship's been hit.' Once a sailor, as they say.

The water had now reached the lower levels, it was running along the corridors, under doors and into the bedrooms, flowing into the canteen, the offices and the recreation room and fusing lights as it went. Worst of all, the Radio Room was starting to get seriously wet and we were in great danger of losing communication with the outside world.

By this rime, various people were beginning to get to grips with the situation and someone struggled into the Control Room and switched off the pump. Electricians began isolating equipment and the Walter was leading a rescue party to save the radio equipment. Slowly, the water stopped flowing down the stairs. There was a growing realisation that the night shift was not going to get any dinner.

In the meantime Jim, the diver and I, having solved our problem, still blissfully ignorant of Armageddon upstairs, decided to have a well-earned cup of tea and some chips in the canteen. As we made our way onto the main deck we were splashed with water dripping down the outside of the accommodation; Jim said, 'I didn't think it was supposed to rain tonight.' Another wrong conclusion based on facts.

What really hurt was, even though we had done a good job, we didn't get any chips.

The clear up operation started in earnest at first light and it was good to see how hard everyone worked to get us back to normal. The stewards threw out all the carpet tiles, mopped floors for hours and threw away sodden bedding. The drilling rig had its own generators, so we ran a temporary supply cable to the accommodation, which enabled us to couple up the deep fat fryer.

Having—thankfully—saved the radio room from immersion, we contacted all the Rigs and Installations in the area to beg portable heaters, bedding, towels and dry overalls for those with soaked clothes. In the meantime, we had something of a struggle stopping Al from finding and then killing the Control Room Operator.

And I guess with some justification. Al said that had this happened on a ship where he was responsible for discipline, he would have keel-hauled the bastard'

Just another day at the office really.

The next 'interesting' episode was even more surreal.

I was summoned to a meeting in Vince's office, there to be met by our four Senior Construction Engineers and Walter, looking even more pensive than usual. You could see the unspoken question on everyone's face, 'What does the evil sod want now?' However, we were soon to be enlightened.

Apparently, the local newspaper had received a muffled phone call saying that a bomb had been planted on board and was set to explode at twelve noon the following day. The Editor had contacted the police who in turn, informed our company. This meeting was called to 'discuss the problem in confidence' which was a euphemism for figuring out what the hell to do.

It is useful to set the scene with regard to what was happening at the time. The Russians had moved out of the North Sea and gone to invade Afghanistan (another example of the perils of hasty decisions) the Iranian Embassy was under siege, and peace had most definitely not broken out in Northern Ireland.

Given the fact that terrorists were alive and well, collecting fertiliser and displaying an inclination to blow everything up, we obviously took the threat seriously. To the extent that we started to re-write our CV's.

It surely can't have helped that one of the (worst) films going the rounds at the time was called 'North Sea Hijack', starring Roger Moore, that master of the facial expression. He played the hero, a sort of upper class SAS figure named, (you won't believe this), Rufus Excalibur Ffolkes, (I told you.) Anthony Perkins played the baddy, or terrorist as he was known, to his everlasting shame. The idea was that the baddies would climb up the leg of a drilling rig, capture the crew and set explosives to blow the thing up, if a woman playing Mrs. Thatcher didn't do as she was told.

Enter Roger Moore.

Using guile, two or three facial expressions and instructions from James Mason, Moore saved the rig from oblivion. They don't make films like that any more.

I once had the dubious honour of meeting two genuine SAS men; they had in fact carried out an exercise on our platform and were gathered with their boss in Walter's office. The first thing I noticed was that they bore no resemblance to Roger Moore whatsoever. Both were small, skinny, rat-like fellows whose idea of erudite conversation was the odd monosyllabic grunt, accompanied by stares of what I took to be murderous intent. After a minute or two, I gave up the idea of asking how they enjoyed their job. Anyway, back to the bomb.

The plan, as explained by Vince, was simple; we were to brief our respective teams and acting very discreetly, would commence a detailed search of the platform, working throughout the night. We would report back at regular intervals and Walter would update the office and the police.

I remember one of the lads asking what the hell was it we were searching for and what would we do if we found it. At this point the tone of the meeting deteriorated to some extent and Vince, who saw the prospect of a commendation from a grateful nation slipping from his grasp, brought the meeting to order. We were each given a section of the platform to search and, with a final caution to keep a low profile; off we went in our new role as amateur bomb disposal experts.

Now what I hadn't revealed to the others was the fact that I already had considerable experience with bombs, something I was convinced would stand me in good stead one day. I was about eight years old at the time and attended a school on the outskirts of Liverpool. One morning as usual, we urchins turned up for lessons with Mrs Corrigan only to find that our school had mysteriously disappeared, and in its place was a heap of bricks and some smouldering timber.

We were, to put it mildly, nonplussed. We stared at each other—who had demolished the building? How could we thank them? The general consensus was that if we had any sweets we would have cheerfully given them to our unknown benefactor. Then one lad, who knew these things said, 'It was Hitler.'

It's not that we didn't believe him; after all he was top of the class at spelling. What we couldn't understand was how Hitler had known where our school was. Still it didn't seem to matter, as by that time those lucky enough to be wearing wellies were already clambering happily over the ruins. The other thing I couldn't quite grasp was the fact that my parents didn't seem to be keen on Hitler—still that's parents for you.

The only problem with Vince's search plan was that our platform weighed 327,000 tonnes, had four concrete legs with lifts and vertical ladders into the depths, there were thirteen modules all packed with equipment, a massive three-storey Living Quarters, Additional Living Quarters and an Extension to the Additional Living Quarters, assorted offices, welding shops, a drilling rig and spare parts stores.

You begin to get the picture—we are being discreet, we don't know what the bomb looks like and we don't know where the hell it is; and meanwhile, the rest of the people on board are staring suspiciously at a group of Engineers peering into nooks and crannies and trying hard to look carefree.

During the course of a long frustrating night, we discovered clandestine beds, hidden tools, packs of cigarettes, discarded overalls, beer making equipment, about 2700 Playboy magazines (which obviously we didn't look at), a bowl of fermenting orange peel—but no bomb.

As we returned to report progress, we could see that Vince was beginning to suffer from mixed emotions: relief that we hadn't found anything resembling a bomb, but on the other hand what if had we missed the damn thing?

Came the morning and we were all completely knackered, no bomb and a growing realisation of the impossibility of searching every space on the installation. Vince, in true leadership mode had remained in his office all night, pausing only to drink copious amounts of coffee and to colour in sections of the platform considered to be bomb free on a huge drawing. They also serve, etc.

Meanwhile Walter, acting under instructions from onshore, was making arrangements for all non-essential personnel to go over the bridge to the semi-submersible rig alongside. The idea was that at about eleven am, they would raise the bridge and move off to a safe distance in case the worst happened. We found this willingness on their part to watch events from a distance to be very reassuring.

Soon after the evacuation, the platform became very quiet; there were just seven of us left on board, Walter, Vince, four Engineers and me. We gathered in the recreation room and began to play rather desultory games of snooker. Gallows humour prevailed and centred on the reluctance of our Onshore Masters, the police and the Bomb Disposal Squad to share the experience with us.

Vince, in an effort to make conversation and lighten the mood said, 'Did you see the damage the IRA bomb did to Lord Mountbatten's boat?'

It was about this time we began to question the wisdom of being there; what should we do if the thing went off? Would the platform tip over? Was this part of our job description? Whose bright idea was this anyway? Should we wear life-jackets? These and other daft questions were posed and left unanswered. Even Vince was somewhat introspective.

And then it was twelve noon. We listened carefully in case the bang was muffled and kept the recreation room door open for a quick getaway. Then, after what seemed an age, it was ten past twelve, then hours later, it was half past twelve and nothing had happened.

Later that day, when everyone was back on the platform, Walter called us into his office and told us the police had arrested the caller. He turned out to be a disgruntled contractor who had recently been sacked for laziness and thought the hoax call would be a fun way of seeking revenge.

I asked, 'Did they know this before twelve noon?' to which Walter replied he believed so, but they felt it prudent let events take their course just to make sure. This seemed to us to be an admirable illustration of our real value to the Company.

Vince, ever the sage, said 'I didn't really believe there was a bomb on board anyway,' and one of our gallant band replied with some feeling, 'You can believe what you bloody well like now and looking back, I don't remember being asked to volunteer.'

There is nothing like hindsight for making one realise that one has been conned by masters. Again. Even Vince looked a bit sheepish. Not a pretty sight.

22

A Need to Know Basis

There are times when a search for the truth is not always in the best interests of those being interrogated. The adage 'Never ask a question to which you don't know the answer' is good advice. That discretion was the better part of not making a prat of yourself was illustrated at a de-brief meeting I attended onshore.

If you cast your minds back you may recall that on one particular trip I had completed a fairly hairy job on the Buoy. Assisted as usual by the two musketeers and a rigging crew, we removed a damaged part of the oil delivery hose to the tanker and replaced it with a completely new section.

This was a ten-inch diameter flexible hose, with a fabric outer cover and a spiral wound steel inner core. One end of the hose had an extremely heavy bull-nose connector and the other end a flange for bolting to the next section. Each section was some twenty feet long (6.15 metres for younger readers) and weighed about 400 lbs (180 kgs for those born fairly recently).

The new hose was slung under Allan's Bolkow and deposited on our helideck. All we had to do then was pull the old hose off the reel with the aid of our supply boat, lock the reel off and disconnect the flange. The damaged section was then winched onto the deck of the boat and taken onshore. We then rigged up the new section, coupled it to the existing hose and test the joint. It's called work.

Nothing to it, job done, off for a brew and mutual congratulations.

Let me move the clock forward to the de-briefing in the Chief Engineer's office ashore; he wanted chapter and verse on how we had managed the job in such a remote location without losing too much in the way of production. After I explained he congratulated my team and me and asked me to write up a procedure so that others could follow suit, if ever required to.

I was just basking in a warm glow, thinking modestly of a salary rise or an article in the Company Magazine, when a young engineer who was attending the meeting asked, 'What did you do with the oil in the hose?'

'What?' I said, with visions of glory fading fast.

However, before I could answer, the Boss gave our questioner a sort of withering look and said, 'Just nip out will you and sit in my office, I'm expecting a call from the States at any time now.'

The lad said, 'But what about the meeting?'

The Boss replied, 'Never you mind about that, we'll finish off here'.

As the lad reluctantly made his way, blissfully unaware that he had nearly caused us both to have heart attacks, the Boss looked at me and chuckled, 'The naive bastard, where the hell does he think it went?' He shook his head. 'No, for god's sake don't tell me.'

I answered, 'I've no intention of telling you, but let's just say the Skipper of our supply boat was not best pleased when the hose swung across onto his deck.'

With that he gave me a fearsome grin and said, 'So you didn't tell him either? Good lad, come on let's go and get a nice cup of tea.'

'But what about your call from the States?'

'What bloody call?'

And off we went to the canteen.

23

The Male of the Species

The first four or so years I spent offshore were in an all-male province. The accommodation had been designed with chaps in mind and there were no separate facilities for the opposite sex. The majority of the cabins contained four berths with two-tier bunks on each side and a toilet and shower in an alcove. The major obstacle to the provision of female facilities was simply lack of space. Not only was every permanent cabin filled with hairy individuals, but also we usually had either temporary accommodation on board or there was a Semi-Sub alongside, similarly over-subscribed in the sleeping department.

So here we were, in an all male preserve, eating ice cream without being told off and pretending to be misogynists although we really had no idea what the word meant. Most of us were only aware of it because Germaine Greer was in the habit of addressing the male population in this way. In fact, it would be more accurate to describe some of our compatriots as misanthropes, as they gave every impression of disliking people *in general*. No discrimination as far as they were concerned.

Things began to change for us at the time we were commissioning the Main Oil Export System and were having considerable problems with the high-pressure pipework. Despite our best efforts it became clear we were unable to solve the problem, a situation that forced us to do something akin to professional suicide—admit we would have to seek expert help from onshore.

I may have mentioned before that we had a healthy dislike of 'that lot onshore' and the thought of them having to come to our rescue was hard to swallow. Even Vince questioned me at length regarding the advisability of taking such a controversial step. The problem was that both he and I were in a cleft stick; each day that passed without being able to run the export system jeopardised the completion programme, our jobs and the chance for the Company to earn lots of money. There was nothing for it but to ask, but in a way that suggested we were actually doing 'them onshore' a favour.

I explained our (minor) problem to my onshore contact and he told me he would arrange for expert help to be made available. Sure enough, some hours later I received a telex saying that a Senior Piping Engineer would be travelling from London on the first available flight and the name on the manifest would be Julie Thompson.

In panic, I telexed back, 'Thanks for the speedy reply, but there's been a misprint, I guess it should be Johnny Thompson.'

'No,' came the reply, 'It's not a mistake, it's definitely Julie; she's a highly qualified Piping Engineer. Good luck.'

I broke the news to Vince and the lads and was inundated with offers to share their cabin. Even though this would be a burden, they felt it to be their duty, no sacrifice too great they said—and other assorted rubbish.

I said, 'Someone has to give up his cabin, not bloody share it,' at which they muttered, 'Ungrateful sod.'

As accommodating Julie safely was beyond my managerial skills, I tentatively approached Admin Al, explaining that all I needed was an empty cabin, preferably at the end of a corridor, with a lock on the door, nice new towels, decent soap, some coat hangers and cleared of any old copies of *Playboy*.

I won't go into details regarding Al's immediate reaction but he clearly thought I had gone barking mad. However, and true to form, he rang me later to say a cabin was available and reminded me that I would be forever in his debt.

The flight bearing our first lady was due just before dinner and before it arrived fighting broke out over who would meet her off the helicopter. Much to everyone's disgust, Walter pulled rank, saying that as the OIM it was his duty to do the honours.

Now throughout history us men have been trying to impress women and failing. However we keep on trying. Remember that up till then no one had any idea how old our visitor was, what she looked like, whether she was married or whether any of the idiots currently foaming at the mouth would be of the least interest to her.

What, of course, we had failed to consider in our haste to play Sir Walter Raleigh, was that she was well used to assorted gorillas preening and fawning, and being much cleverer than us, remaining singularly unimpressed.

Having been made officially welcome—patently ignoring the serried ranks of blokes accidentally lining the corridor—she had dinner with Walter, Vince and I, during the course of which we outlined the problem. As it was then fairly late, we agreed she should meet in our office at seven am to review the pipework calculations. After which, she bade us good night and retired to her cabin to prepare for a full and exciting day on the morrow.

On entering our office at about quarter to seven the next morning, I was struck by several significant changes. Everyone was early, everyone was wearing a clean shirt, everyone's hair was combed, everyone was sitting at a tidy desk and the Pirelli Calendar Girls, who normally adorned the walls, were missing.

That of itself was fairly revolutionary, but was insignificant when compared to the smell.

It was so powerful you could almost see a cloud hovering above each Engineer and Technician. There was an overpowering mixture of Brut, Old Spice, Gropers Delight, Denim, Dynamic Pulse, Police, Team Force, Jovan Musk, Polo, Baron, Sex Appeal, Liniment and Aramis, to name but a few.

You may wonder how I knew the names, which you have to admit were a wonderful mixture of masculine and naff. I will tell you later.

Choking in the fumes, I accused the gang of preening to impress our visitor. To a smelly man, they vehemently denied any such thing and professed they had no idea what I was talking about.

Just as the denials were dying away, Admin Al walked in, stopped dead and cried, 'Christ Almighty, it smells like a Turkish belly-dancing contest in here.' Even though I couldn't confirm his statement (having lead a sheltered life,) I felt vindicated, despite the assorted mutterings for Al to 'sod off and mind his own business.'

Fortunately, by the time Julie turned up to start the investigation I had turned the extract fans up to maximum and contrary to all safety requirements, wedged the door open for ten minutes. Slowly the worst of the haze began to dissipate and we were left with a manageable residue, not unlike the perfumery section in Boots.

Later that day, Laura and the piping team were busy trying to resolve our technical problems. I had sent some telexes ashore and was chatting with Walter in his office when Larry, the Camp Boss came in. He asked how things were going and I related my tale about the lads trying to make themselves irresistible to our guest.

Larry said, 'Bloody hell, now it makes sense.'

We asked him what he meant. 'Well, I couldn't for the life of me figure out what was going on last evening. When I opened the 'duty free', there was a huge queue and everyone wanted after-shave. So much so, I sold all my recent stock and then got rid of a load of rubbish I'd had in the back room since we sailed out of Norway three years ago.'

Apparently, over the years, he had accumulated a vast selection of stuff that no one wanted and he reeled off a selection of the good the bad and the ugly of the after shave world.

The penny dropped - now the overpowering smell started to make sense. Walter said, 'Did you manage to satisfy everyone?'

'I sure did,' replied Larry, 'They got so desperate towards the end, that one daft sod asked whether there were any ladies' perfumes when I told him we had sold out of men's stuff.'

We leaned forward in anticipation and not a little alarm to hear the outcome.

Larry continued, 'I told him I've got plenty of women's stuff for sale, it depends whether you want to frighten your wife into thinking you've gone on the change, or risk being chased around the platform by some short-sighted hairy welder.

So he asked me what else I had. I said I had some liniment in a fancy bottle; it contains Wintergreen, which I am told, is a smell that's irresistible to women. The idiot took it.'

Walter said, 'It just goes to show that man's pathetic attempts to impress the unimpressible are alive and well and failing in a grand manner.'

Larry and I nodded in agreement, although we hadn't really understood what he was on about. But he was after all, the OIM, and a lot wiser than us.

24

Fifty Knot Fog

For the benefit of any readers living, for some obscure reason, south of Hadrian's Wall, a word of explanation regarding the Shetland Isles. They are located in the North Atlantic about 150 miles north of Aberdeen (240 kilometres for younger readers), they are higher up than Moscow and the people are prone to using strange sounding place names like 'Muckle Flugga.' With the North Sea on one side and the Atlantic on the other, they are remote, stark, windy, wet, treeless, beautiful and prone to fog.

We're not talking here about your average, run of the mill, softy, girly, velvety fog; this stuff is heavy, wet, thick, evil, long lasting and when provoked, can travel at high speed. When it arrives it feels it's only fair to hang around for a while, say four or five days. There is nothing wishy-washy about Shetlands fog and there is nothing indecisive or wispy in the manner of its arrival. It simply descends, spreads out over the islands and the surrounding North Sea, so that roads, houses, sheep, helicopters, massive offshore platforms, daylight and small children simply disappear. In other words this is strong, manly, Pavarotti type fog made from girders and is definitely not to be confused with mist.

Are you getting the picture? (No pun intended).

My reason for the lengthy, but educational explanation is that I need to set the scene for the next breathtaking episode—Crew Change.

Crew change in the Central North Sea was a fairly straightforward matter. You went to Aberdeen Airport, registered your presence, had a safety briefing, boarded a helicopter and an hour and a half later you were on the platform. There were a few delays, but in general they were due to high winds offshore.

Crew change in the Northern Sector required the introduction of another element, that of a fixed-wing flight from Aberdeen to Sumburgh Airport,

185

at the southern end of the Shetlands. You then disembarked, registered your presence, went through a safety briefing and boarded a helicopter to the platform. Or not, as the case may be.

It is no bad thing to be paranoid as long as you can justify your paranoia. In the case of crew change, it was everyone's firm belief that whatever the day allocated for crew change, the fog would arrive just as they were about to travel. This applied whether you were going to, or coming from the platform, but with two key differences.

Being thwarted when setting off to work was largely inconvenient and extremely boring; you would arrive at the airport at some ungodly hour in the morning to find that all flights were delayed. This state of affairs could last until about four pm when the Company Representative would inform everyone that there was to be no flying that day (as if we hadn't worked this out for ourselves.) Confirmation of our worst fears generally led to grumbles all round, and the problem of finding a hotel bed for the night, ready to try again the next day.

The next day the process was repeated, arrive at the airport, find a seat, buy a newspaper, a packet of mintoes and wait for something to happen. Sometimes you were given a voucher to stave off starvation; the only drawback was that it had to be spent at the airport Café. This somehow neglected that they had only laid on sandwiches for the normal number of clientele and not for some two hundred hairy, bad tempered offshore workers, all of whom felt entitled to a proper three course dinner, including chips.

My record for this stimulating lunacy was three consecutive days, after which my brain was in some kind of advanced shutdown and I would have flown anywhere just to relieve the boredom.

This trauma however, paled into insignificance when compared to being thwarted when trying to get *off* the platform, a non-event that raised one's levels of rage and frustration to undreamed of heights.

Picture the scene; you have carefully prepared for disembarkation. Starting the night before, handover notes were written, taxi requirements were telexed onshore, perfume and cigars were purchased from the Camp Boss, your collection of double sided cassette tapes featuring such *avant-garde* acts as The Jackson Five, the Osmonds and The Ramones were carefully packed away ready for your next trip.

On the morning of departure, you rose early, ate a good breakfast, showered extra carefully and put on your 'going home' gear. You checked the expected arrival time of the crew change chopper and you made your way to the departure lounge.

Mind you, 'departure lounge' was what it was *called;* any resemblance to an actual facility of that name was illusory. Our lounge was located in a corner of the top floor corridor and was furnished with about six chairs deemed too dangerous to remain in the canteen. At the beginning, some bright spark thought it would be a good idea to install a tropical fish tank in the corner, the idea being to soothe and cosset us prior to take-off. So with a great deal of effort, a tank was glued together and the budding naturalist brought out a selection of fish to be installed in the new aquarium. To complete the picture, there was a plastic sunken pirate ship, a rock, an aeration pump and some weed.

This arrangement worked well throughout the summer months right up till we had our first autumn storm. Remember, the departure lounge was located at the highest point on the platform and when your average sixty-foot wave hit the concrete legs the whole installation shuddered, which unfortunately created a mini tsunami in the fish tank.

As the wave raced from one end of the tank to the other, the fish found themselves involved in situation reminiscent of surfers at Bondi beach.

All was okay until the wave reached the end of the tank where, due to its momentum, the water carried on over the lip of the tank and deposited itself and most of the fish on the floor. At this turn of events, Al looked up from his flight lists and with a withering glance at the devastation said, 'bloody fish.' Once a sailor as they say.

Funnily enough, I noticed that on my next trip, the tank had vanished and been replaced by a two more plastic chairs and an unframed picture of a hunting scene.

Back to the plot. Spirits were still high, plans to avoid wallpapering the lounge were being firmed up, whippets were to be exercised, and car showrooms were to be visited. And then...

Signs begin to emerge indicating that things were not going according to plan. The Helideck Crew were not mobilised, the Steward with sandwiches for the Pilots had not appeared, the mail bag had not been brought up and then, Admin Al came out of his office and said, 'Right you lot, there's thick fog on the Shetlands so you might as well piss off back to work.'

Hatred and disbelief are a terrible mix, especially when you don't know where to direct the emotions. One thing was certain; nobody dared direct his or her frustrations at Al.

But go back to work?

A variation on the above scenario was to get up on the longed-for morning to find that you couldn't see anything. The whole platform was engulfed in thick white stuff that wasn't there when you went to bed the night before. Where the hell had it come from and, more importantly, how long would it last?

Down you went to the Radio Room to get the latest weather forecast from the Met Office. Lots of incomprehensible rubbish about warm fronts, troughs of low pressure, relative humidity, precipitation and seven-day predictions. The same crucial, unanswered questions—'Will the fog disappear in the next ten minutes and if not, who can we blame?'

And always there was the unspoken fear that Admin Al would broadcast another of his succinct statements regarding a return to work. As you can see, going-home delays were far more frustrating than coming-out delays. Carefully laid plans were jeopardised, especially for the unfortunate souls who had arranged to take the family to Majorca. Visions of the family with buckets and spades packed, waiting for dad to arrive in time for the charter flight that afternoon soon began to haunt them.

As one guy said, 'The wife will kill me if I don't get home in time, she's been going to a Beauty Parlour for three weeks to get a nice tan and she definitely won't want to hear that a bit of mist has deprived her of the chance to look like Carmen Miranda.'

There is a myth about fog that it is still and tends to lie around like a teenager. This may be the case onshore, but *our* fog was intent on going places. On a number of occasions we went up to the helideck to find the damn stuff racing past at a rate of knots. Newcomers to this phenomenon would be quite cheered, thinking, lovely, this wind will blow it all away in a minute or two. What they had failed to grasp was that the North Sea was full of fog and even at the magical fifty knots it would take about two days for it all to whizz past us.

Another really evil trick nature could play was having the fog clear enough for the crew change chopper to set off, only to find that although the sky was an azure blue when flying at two thousand feet (575 metres), our helideck was still completely invisible. So we heard longed-for sound of a helicopter making the lovely thrum, thrum sound, just like Robert Duval's in 'Apocalypse Now', (but without the 'Ride of the Valkyries',) and then the pilot's voice telling Admin Al that 'As he couldn't see where to land, he would have to return to the Shetlands.'

Now I realise you are already aware that us offshore chaps were made in the mould of John Wayne when he played the part of Red Adair in the film *Hellfighters'*. However, when we heard that lovely 'thrum thrum' disappearing into the gloom, stiff upper lips could be seen to wobble as we realised that having turned back, the pilot would now be 'out of hours', the chopper would need servicing, those aboard—on their way to work— would be rioting and it would be dark in two hours.

As Walter commented, 'It's a good job we joined for the excitement.' Actually, that wasn't *exactly* what he said, but this is a family book.

As an aside, you may also have noticed by now that I have a tendency to relate major events to films. It can't be helped; I had a misspent youth, going to the cinema about twice a week, during the course of which I memorised not only the titles and stars but also the supporting players, the directors and the musical composers. Even now I know that Saul Bass created the opening credits for 'The Man with the Golden Arm?' (Frank Sinatra going cold turkey.) If you don't believe me, look it up. The problem is that this fantasy world has been apt to creep into my everyday life, leading to some pitying looks from anyone on the receiving end of all this Barry Norman information. Still, I didn't smoke.

Back to the fog--incidentally 'The Fog,' released in 1980, starring Jamie Lee Curtis, was a great film as well, as was Jamie Lee Curtis.

Now, thanks to the prevailing conditions, we had two groups of people one hundred miles apart, each in a total slough of despond.

Those on the platform realised that the special shower and liberal use of after-shave had been wasted and that overalls would be the sartorial choice for a little longer, whilst those trapped in the Shetlands Airport realised they were going nowhere and the exciting prospect of finding non-existent accommodation would occupy them for the next three hours.

On a more serious note, I remember on one occasion being trapped in the airport in similar circumstances and got fed up with sitting on the tiled floor with my back to the wall. So to relieve the boredom, I staggered up and met one of the airport staff whom I vaguely knew. I began the usual tirade against the fog, thinking he would sympathise, when his reply took the wind out of my sails, so to speak.

'We never curse the fog; in fact it saved many lives during the war. Our real fear, when we sailed in and out of occupied Norway in small fishing boats to carry out sabotage against the Germans, was moonlight. We prayed for fog, as it was the only way we could sneak in and out of the Fjords; we called these trips The Shetland Bus.'

I suddenly felt humbled and asked my friend to tell me a little more about this mysterious wartime enterprise, which he did.

I cannot possibly do justice to his story of daring and immense bravery, so I will just remind you that it is some two hundred miles from the Shetlands to Norway and you have to cross some of the most dangerous waters in Europe.

He told me about one particular Norwegian named Leif Andreas Larson who made some fifty trips to Norway and was awarded the Distinguished Service Order, Distinguished Service Cross, Conspicuous Gallantry Medal and Distinguished Service Medal and Bar. The volunteers were paid £4 a week with free food and lodging and a bonus of £10 for every trip they made to Norway. (Obviously no expense spared there then.)

I realised whilst listening to the stories, that it was one of the best-kept secrets of the war. Since then, I made it my business to read about the Shetland Bus and I urge you to do the same. It will bring a whole new meaning to the word 'bravery'. (For a good start try www.shetland-heritage.co.uk/shetlandbus/).

Talking of Shetlands accommodation, sometime in 1979 I was the reluctant recipient of this five-star hospitality. This time it wasn't fog but a blizzard that had blown straight down from the Arctic and, seeing buildings for the first time, decided to land in great heaps on the runway at Sumburgh.

Unfortunately, being largely indiscriminate, it also landed on the Viscount aircraft that was supposed to take us back to sunny Aberdeen.

After peering out of the windows watching the landing crew trying to de-ice the aircraft, clear the runway, stand upright and avoid hypothermia, we realised it was hopeless, we were going nowhere. Darkness was settling in nicely, the wind was now so strong it was blowing the snow into the sea and the Viscount's pilot and flight crew were nowhere to be seen.

Now, although there were only twelve of us from our platform, the Viscount carried some eighty passengers, which meant we were joined by crews from many other offshore platforms and rigs. And all of us were by now sharing the same growing feeling of unease that we were stuffed.

But what made this even more obvious to the sharper intellects among us was the sight of the airport staff making rapid preparations to blow out the landing light candles, shut up the café facilities, lock the doors and go home for the night. We were soon united in a common hatred for these lucky sods with nice warm homes to go to—always assuming they could find them under the snow.

It was now time to remonstrate with the Company Representative, an unfortunate individual who, in a moment of weakness, had been persuaded by some sadist in the Logistics Department to live a nomadic existence at the Airport. He had the enviable task of seeing us on our way, either travelling out to the platform, or travelling to Aberdeen—or in this case, not travelling to Aberdeen.

He was now cornered in his little office by a bunch of somewhat less than enchanted chaps, who, having reluctantly de-planed, were now demanding to know what he intended to do about accommodation for the night and could he get a bloody move on, as we were sick of the Company, the weather, the Shetlands, each other and softy Airline Pilots.

The Company Rep had noticed by now that reason and logic were not high on the list of priorities for the malcontents gathered around him, a number of who were just about to climb over his desk. However, as one of the main strengths of a Company Rep is the ability to lie with confidence, and in fear of his life, he played what he considered to be an inspirational trump card.

'Don't worry lads; I have arranged accommodation for you all at the Sugbay Hotel.'

Not being familiar with this establishment there was a collective gasp; the threat level went from 'immediate' to 'stand-by' and some idiots almost applauded.

We knew the hotel wasn't very far away as, on a good day, it was visible from the aircraft when we came into land. Spirits rose and discussions containing key words such as fillet steak, malt whisky, kip, chips and sticky toffee pudding began to be heard.

'Right lads,' said the Rep, who was still amazed that we had swallowed his imaginative solution, 'Your transport's outside, let's get you off to the hotel.'

Of course, in our euphoria, we had forgotten one small but important fact. At the same time as we were being lied to by the best Company Rep in the world, the same wonderful news was being proclaimed to about sixty-eight other hairy chaps by *their* Company Reps.

So, what began as a gentle stroll to the coach became the sort of riot similar to that were seen in India when they decided to leave the Empire.

In a single heaving mass, eighty guys plus bags, cases and duty free fags, tried to get through a swing door and onto a coach with luxury seating for twenty-four. You had to admire the sense of common purpose. Finally after moving five blokes from the drivers seat, we were all on board; and with the bus bearing a marked resemblance to the last helicopter out of Saigon, we set off for the short run to our new destination.

On our arrival, we began to wonder just how the hotel, which looked for all the world like a medium sized country house, could provide beds for eighty guests.

We needn't have worried—well actually, we should have.

It turned out that our accommodation was not actually in the country house part of the hotel but in some huts in the grounds. These days, a Holiday Brochure would probably refer to such a layout as—'The Annex, set in its own grounds, delightfully secluded, but within easy reach of the main complex.' But of course, we didn't know you could describe huts like that at the time.

So when I say huts, I use the word in its broadest sense and the only way I can convey the awful sight that awaited us is to revert to films (again) as a comparison. Sorry about that.

Even my very young readers must be familiar with a film which appears on TV every Christmas called 'The Great Escape.' Therefore you will remember how the prisoners lived in rows of wooden huts with bunk beds, rotten windows and very basic ablution facilities. The similarity of the film set to our accommodation was staggering, right down to the lack of curtains on the windows and the fact that there was only cold water coming out of the taps. (Remember that outside, the blizzard was alive and well and intent on coming inside).

Anyway, in the end we did get chips, we didn't get fillet steak, we did get some beer, we didn't get malt whisky, we didn't get a nice warm shower but we did get some kip, but only by wearing all our offshore thermals.

As I have always said, 'You can't do enough for a good Company.' Oh, and rumours that our esteemed Rep had asked for an immediate transfer were false. When we got back to the airport, we found he had simply gone to his Grandma's funeral.

25

Problems, What Problems?

Somebody once said, 'Don't tell your problems to other people, 80% don't care and the other 20% are glad you have them.'

I suppose we had our share of problems; some were fairly serious but the vast majority were idiotic. You will notice my continued refusal to call them 'opportunities'. The following are examples of the adage, 'Bad planning on your part does not necessarily constitute a problem on my part.'

A Communication Problem

There was a postscript to the Sugbay Hilton saga, which came to light on our next trip. Apparently Vince, who was with us at the time, hadn't been able to get his usual eight hours of uninterrupted kip and so lay festering throughout the night. The following morning after a full Scottish breakfast consisting of a round of toast and some lukewarm tea, (taken in the main complex,) he demanded to see the owner of the establishment.

The lady who was nominally in charge tried to fob Vince off with the usual promise to 'be sure and let the owner know that you were disappointed'.

'Disappointed?' yelled Vince, 'I'm not disappointed you silly woman, I'm bloody well enraged and I want the sod here, now, so that I can assure him that my Company will never send anyone here ever again. Just you fetch him right now so I can convey to him in words of one syllable the problems with his appalling establishment.'

This was Vince at his vintage best and something we were well used to, as we all bore the brunt of his gift for understatement at regular intervals.

Unfortunately, the lady hadn't met any red faced, ranting lunatics like

Vince before, especially not while serving breakfast. Again she made a fatal error and began to try and reason with Vince. Sadly we could have told her that as a placatory strategy, this was doomed to failure.

Vince predictably went into orbit. 'Get me the bloody owner now or I'll get my lads to search the place until we find him and I can sort the miserable con-man out, he aught to be in bloody Craiginches for conning innocent workers like this.'

Even though we had once again been recruited without consultation, we rather liked the part about Craiginches (a very old and very crowded prison in Aberdeen.)

The lady, who was now quite close to tears, said, 'Don't you swear at me like that, I am doing my best to explain the situation.'

At this, Vince looked somewhat perplexed, as he didn't class what he was saying as swearing, and had one more try. 'Madam, will you please bring the blasted owner to me—*now!*' This last word spoken at about 130 decibels.

The lady stared at him with a mixture of fear, hatred and exasperation and in a quavering voice said, 'I keep telling you, you horrible man, I can't get him because he always spends the winter in the Bahamas.'

Biggles Lives

Staying with flying for a minute, I had an interesting flight in 1976 due to a problem with British Airways (unlikely, I know.) I turned up one morning at Manchester Airport ready to get the 'red eye' to Aberdeen, which was pre-booked for me by a local travel agent. On reporting to the check-in desk I was informed that my flight was full, but not to worry, they would sort it out later and handed me a £2 food voucher by way of recompense. It turned out that three of us Aberdeen passengers were victims of 'over booking', a caring strategy the airlines were prone to instigate in those days.

Their rationale was based on the iffy premise that not everyone who had booked would turn up and consequently, they would be left with empty seats, thereby reducing profits by 0.0001% on each flight. The reality was of course, that on early flights, everyone did turn up and in the ensuing lottery, those who didn't check in soon enough would be left behind. It's known in the trade as customer service.

Sadly, as far as the airline was concerned, the three of us weren't too happy with their novel accounting practices. I realise this may sound unreasonable, but that's passengers for you; they have an unrealistic expectation that having booked a seat, they should be allowed to fill it.

The next logical step was to conduct a rather undignified shouting match with the 'passenger liaison' person; who, having being pinned against the wall, uttered those immortal words, 'Leave it with me and I'll see what I can do.'

As a result of our agreeing to this rash promise, we were led out to a tiny 'Cessna type' aeroplane. Which, as proof of its intention to go somewhere, was revving its engine to such an extent, that the whole thing was threshing about like a demented wasp.

I was reminded of the doubtful reassurance given to passengers in Dakotas during the war. 'Don't worry, when an engine fails on a twin engine aircraft you always have enough power left to get you to the scene of the crash.' Not too reassuring in our case, as there was clearly only one engine.

Realising that this was the literal translation of 'I'll see what I can do,' we were ushered aboard and I was directed to sit in what I supposed to be the Co-Pilots seat. As I sat staring out of the windscreen, the pilot, who was busy clicking about ninety switches on and off, passed me a pair of earphones and motioned for me to put them on.

Having done so, the frantic engine noise faded to about four hundred decibels and I heard the pilot say 'Can you hear me okay?'

I'd seen *The Dam Busters* (Richard Todd as Wing Commander Guy Gibson) and remembered magic phrases such as 'wilco', 'angels one five' and 'bandits at four o'clock'. I thought, would they be useful now, and should I let him know I was familiar with flying language?

However, as he hadn't used any of the 'jargon' as we in the aeronautics business call it, I confined myself to 'yes'.

'Good,' he said, 'I am going to need your help; the problem is I don't have a co-pilot and could do with some assistance.'

I had an immediate nightmare vision of being asked to take over the controls in mid flight and was about to let him know I would rather not, when he handed me an *AA Road Map of Great Britain* and said, 'I want to follow the M6 and A74 to Glasgow, can you find the page and keep your eye out to see where we are going?

I will be flying at about 8,000 feet and as it's good visibility, you shouldn't have any trouble, okay, thanks.'

And without waiting for a reply, we revved up even more and took off.

Having circled around and presumably pointed the plane north, we were soon over open country with the pilot and me peering out of the front and side windows looking for the M6. What this must have looked like to the two passengers behind is anybody's guess.

Now as this was my first experience of flying in a toy plane, I hadn't realised just how differently they behaved when compared to a Boeing 747. This had an unfortunate side-effect as far as I was concerned because, as each gust of wind blew the plane around like an out of balance shuttlecock, the M6 kept disappearing and then reappearing on the opposite side.

However, as the pilot seemed relaxed about all this threshing about, I began to try and relax my neck, arms, back and leg muscles, which had been locked in a nervous spasm since take-off.

Just as I was slowly unfolding my body, his next co-piloting instruction came through my earphones, 'You'll find a box of chocolates in the pocket at the side, could you open it, have one yourself and pass it round to the other passengers?'

This, I thought, was what Captain W. E. Johns must have been on about when he wrote his *Biggles* stories, one hand on the steering thingy and the other casually dipping into a box of *Quality Street*. Marvellous, and I was beginning to see what a vital role was played by the co-pilot.

We continued to fling ourselves northward and with rapid glances at the map, I reassured him that we were approaching the A74. But on looking ahead, I now began to have another feeling of inadequacy regarding my role as an unpaid flying assistant.

Plucking up courage and trying to make it sound like a casual inquiry, I said, 'What do I do when we get to Edinburgh? There are no dual carriageways to Aberdeen once we cross the Forth. I can see two possible routes but they are very small roads.'

His reply was straight out of *Biggles Air Pioneer*—you know the one where he is outnumbered against the Hun and has to follow the hedges in rural France to get back to Algy and Ginger.

He said, 'That's no problem, you can relax. I'll just follow the coast up to Aberdeen then cut inland to Dyce airport. Are there any chocolates left?'

No wonder we won the Battle of Britain.

The only problem I had when we landed, was to be informed that my £2 voucher wasn't valid in Aberdeen.

Have You Seen Our Pig?

Our largest platform was designed to gather oil from other installations, store it in the base and then pump it to an Oil Terminal on Shetland. To do this we had laid a thirty-six inch diameter pipeline coated in three inches of concrete, 560 feet deep on the seabed. The pipeline was some 96 miles long and capable of delivering about 533,000 barrels every day. (Sorry about the technicalities, but I need to set the scene for the next little problem we encountered).

By the way, 533,000 barrels is about 1.8 million gallons. In real measurement this means that at any one time, the pipe held about 34,000 wheelie bins full of oil—that should give the Council something to think about.

Now while we're still in full technical mode, I need to explain what we had to do before sending the oil on its long voyage. At this time the pipe was full of water, which we had used to make sure the pipe could hold the pressure of the oil when we started pumping.

I mean, imagine the embarrassment if the pipe were to burst when it was full of oil, imagine the loss of money, imagine us all being sacked.

After we tested the pipe, what we had to do next was to flush it out and make sure there was no rubbish such as nuts and bolts, welding rods or dead animals left inside by the construction department (accidentally of course.)

Enter the 'Cleaning Pig.'

This resembles a giant cotton reel made of steel with plastic discs at either end and designed to fit snugly inside the pipe. Sometimes they had wire brushes mounted on springs in the middle of the pig to scrape the walls. I know—when you think about it, the whole concept sounds extremely painful.

The pig we were about to use was about four feet long and weighed about as much as a Fiat Uno. Not something you would want to drop on your toe.

I won't go onto the boring details about how we got it into the pipe, but suffice to say it involved us operating lots of valves, by-passes, lifting gear, safety locks and swearing.

Having got the pig in place, all we needed to do was start the oil export pumps and then using water, push it all the way to the oil terminal on the Shetlands.

Throughout this operation, the terminal would keep a valve at the end of the pipe open and let the water run back into the sea. Obviously, as our construction lads had been meticulous in their clean-up habits, there was no risk of any rubbish been ejected into the sea; or so they told me.

When the pig arrived at its destination, the guys in the terminal would send it into the 'Pig Receiver', which was a by-pass at the end of the pipe, a bit like a railway signalman switching points. Simple really.

As it landed, the pig would hit a link inside the pipe, which would trip a signal on the outside. This consisted of a simple metal disc marked in a bright colour and was known as a 'flag'. Who said we weren't high tech junkies?

Incidentally, our oil export pumps should not be confused with the type you see bubbling in fish tanks; on the contrary, they were gigantic affairs each one being driven by a Rolls Royce RB211 jet engine. The noise of one of these engines starting up was not something you would want to get too close to, but there again, how often do you get your very own jet engines to play with?

I remember speaking to one of the engineers from the pump manufacturers and he told me that when they were testing the pumps in the factory, they didn't have a jet engine so they used a huge electric motor. The only problem was that starting it up it took so much current that a number of areas in the city blacked out. After this had happened a couple of times, the Electricity Board insisted that any future tests took place at night and they would be ready with emergency power switching facilities.

Anyway, being of an optimistic nature, we started a jet engine, sent the pig on its long journey and went for a well-deserved mug of tea and a bun.

The next bit of fun involved a good deal of calculation, mostly on the back of time sheets, envelopes, pay slips and wall-boards. The aim of all this was to determine when the pig would make it to the end of the pipeline. As we were all highly qualified engineers and good at sums we multiplied and divided the length, diameter, flow rate, velocity, temperature and other numbers. The outcome of all this brainpower was a general consensus that the pig's journey would take exactly twenty-four hours, give or take a day or two.

We obviously needed to capitalise on this insider knowledge and we quickly set up a lottery selling tickets for £1 each, the winner to take all.

I remember there were some murmurings of discontent when Walter came over all moral guardian and said he didn't think gambling was allowed offshore. Fortunately we managed to convince him it was for a good cause and set aside a percentage (requiring yet more calculations) to go to the Children's Christmas Party.

All we had to do now was sit back and wait for news of the pig's arrival and pay out the dividend to the lucky winner.

However, disquiet started to set in after about thirty-six hours when there was no news from the oil terminal.

For a time we consoled ourselves with theories regarding debris in the way and the pump not pushing the pig fast enough. What we didn't want to consider was the nightmare scenario that the pig was stuck somewhere along the ninety-six miles of pipe, with the water was simply forcing its way past. This seemed to us to have all the makings of a problem and not, as idiot Guru's would have it, an opportunity. Unless of course, one meant an opportunity to resign.

During this nail-biting period, some fool pinned a picture on my office door of Pope John Paul, who had just been voted in with lots of white smoke. The caption underneath said:

COMMISSIONING DEPARTMENT ASK
NEW POPE TO FIND PIG.

Tasteless really, but I wasn't about to refuse help from whatever source.

By now we were well into the British tradition of blaming one another, but even this diversion didn't seem to be working, as there was *still* no word of the pig.

Remember, this was still the nineteen-seventies; nowadays pigs are what is known as 'intelligent' and have all sorts of clever gubbins which can send signals to tell you exactly where they are. No more frontier stuff for the current, softy breed of offshore tigers.

Not only were we on our own regarding the pig's whereabouts, communication with onshore was still by telephone, which meant that connections had to be made by others, leading to all sorts of frustration and dare I say it, rage.

I was now faced with several dilemmas, how to minimise the loss of what little credibility I had left; what reason, if any, could I invent for pig non-appearance which would satisfy my masters onshore and most important, how could I prevent the new Pope from losing his infallibility status.

There was nothing for it but to try to make contact with someone at the Terminal to find out whether the pig had arrived. This course of action was fraught with danger as I would be dealing with a rival company (the enemy) and an admission that we were unable to resolve our problem ourselves would be like manna from heaven to them.

This healthy state of affairs exists because each oil company is convinced that they and they alone, have the requisite expertise to find and produce oil; and that all other companies are by definition, bungling incompetents.

I had to tread extremely carefully. Was I the right man for the job? Could I lie convincingly? Could I persuade anyone else to make the call? Not a chance.

What I had also failed to consider in my excitement was that it was a Bank Holiday and as the terminal was not yet in production, there was a complete absence of rival personnel ready (and willing) to answer the phone.

This minor setback resulted in a protracted series of calls back and forth to the Operator, with requests to try yet another number as each one failed to find anybody. After about half an hour of this I began to feel that I was trying to make contact with the dead and the Operator was beginning to show his frustration, evidenced by protracted sighs and unnecessary comments like 'Oh it's you again.'

Finally, success – after about ten tries, I heard the reassuring sound of a human voice on the other end, speaking with the soft lilt of a Shetlander. At which point, began one of the most surreal conversations I have ever had.

'Hello, is that the terminal?'

'Oh helloo, yes, good day to you. This is the terminal, who is speaking please?'

'It's the platform here; we need to know if our pig has arrived.'

'Helloo, did you say pig?'

'Yes, we haven't had any word from you regarding its arrival and we are getting anxious, could you check for us?'

'Oh well, that could be a problem, as there is no one here at present.'

'Look, I can't stress enough the urgency of the situation, could you check the Pig Receiver and see if the flag is up?'

'Helloo, what was that about a pig receiver, we haven't had any deliveries recently and I know nothing about a pig being delivered.'

'You say there is no one there, who are you?'

'Helloo, my name is Sandy Cooper, I'm the watchman on duty over the holiday and I can't see any flag flying from here.'

'Where are you—are you on the plant?

'Helloo, well almost, I am in the Gatehouse. What is this about a pig, are you playing a joke on me?'

'No, really, it is most urgent and I do need your help, the pig is coming through the pipeline from our platform and it should have arrived by now.'

'Helloo, why would a pig be in the pipeline? Who is this? I think you are playing a joke and I am going to hang up now.'

'No, please don't, I'm speaking from our platform and the pig is a big steel drum which we send through the pipe to clean it. Really I am serious, it's not a joke, please don't hang up.'

'Helloo, alright but I don't know what I can do'

'Thanks, do you know where there is a lot of water is coming out of a big pipe on your plant?'

'Helloo, yes I've seen it on my walkabout, it's at the far end of the road by the Voe.'

'Great. Now Sandy could you go and have a look at a blank pipe next to the open pipe, you should see a big red disc on the top; this is called the flag and will tell us whether the pig has arrived. Could you see if the disc is upright and plainly visible and let me know as soon as you can?'

'Helloo, yes, I will go and have a look and call you back.'

'No, Sandy, you can't do that; I will call you in one hour, thank you very much.'

'Helloo, right ho, bye-bye for now.'

This conversation had lasted about thirty minutes and on putting down the phone, I discovered two things, one I was lathered in sweat and two, our guys were staring at me as though I was mad.

They of course had only been privy to one side of the conversation and couldn't figure out what the hell was going on. So I tried to explain.

'It seems that the fate of our multi-million pound oil export enterprise now rests in the hands of a security guard named Sandy, who half thinks it is someone playing a joke on him. He hasn't an ounce of technical knowledge in his body and I've persuaded him to go onto a plant with miles of pipework to find a flag that he has never seen before.

'But otherwise it's all going to plan.

'All I have to do now is ring him back to get the good news.'

As you can see I was in a sort of hysterically optimistic mode.

The lads looked at one another and it was easy to interpret the unspoken message—what the hell is he on about, how did we get mixed up in this farce and how can we keep it off our CV's?

All too soon it was time to make another call my new friend Sandy. So with what I liked to think of as unwavering support from the lads, I put through the call. Such was the state of my nerves I found Sandy's sing song reply quite reassuring; at least he hadn't got lost or fallen into the sea.

'Helloo, Sandy here, is that the platform calling?'

'Sandy, yes it is, good to hear from you, are you okay, did you find the flag?'

'Helloo, yes I think so, is it a round metal thing, sticking out of the top of a big pipe in a bright red colour?'

'Yesss - Sandy you're a genius and a life saver, are you sure it's in the vertical position?'

'Helloo, oh yes it is, I could clearly see it when I climbed up the ladder at the side. Do you want me to pull it down for you?'

'Sandy, no for God's sake don't touch it, your news is just what we want, many thanks.'

'Helloo, there's just one thing still puzzling me.'

'Oh Sandy, what's worrying you?'

'Helloo, I haven't seen any sign of a pig.'

'No, really, it's okay Sandy, it will be at the end of the pipe just past the flag and don't worry, the terminal crew will take it out when they come back off holiday.'

'Helloo, I think I could get it out if you want me to.'

'Sandy, no for God's sake don't touch the valve at the end, you will injure yourself. Besides the pipe is still pressurised by our pumps and you could be blown clear into the Voe. You've saved our lives, thanks again for your help and goodbye.'

'Helloo, bye-bye for now.'

I sagged back with a mixture of relief and exhaustion and one of the lads said, 'I thought you were going to say he's saved our bacon.'

'Oh very droll,' I replied through gritted teeth, 'Now sod off and turn the bloody pump off.' As you can see I was still my calm fully-in-control self.

On reflection, I began to realise just how helpful this unknown chap had been and I could sense towards the end of his conversation that he really wanted to get a glimpse of the mysterious pig. I suppose it would be just to satisfy himself that he hadn't been the victim of some cruel joke.

There is an interesting postscript to the tale; at the end of the saga, someone added another note to the Pope's picture, which read:

FINAL SCORE: NEW POPE – 1 COMMISSIONING – 0. POPE
FINDS PIG.

Sadly, he was only in office for 33 days when he died, after which another postscript to the message appeared, which read:

Okay, BUT IT BLOODY WELL KILLED HIM.

Who said offshore workers lack sensitivity?

But just to be on the safe side, I decided to remove the picture before I got the blame for any ills that might have befallen his successor.

A Call from the Heart

I had an unusual phone call one morning from the Personnel Department asking if I could locate Tom, one of my technicians, and bring him to the office, as there was a need to contact his wife on a matter of some urgency.

As if to reinforce the delicacy of the situation, the woman on the other end started to talk in hushed tones, to such an extent that I could hardly hear what she was saying. I asked whether there was something I could do to prepare him. Had something dreadful happened to the children? Was she leaving him?

You can see I was now going smoothly into 'agony aunt' mode. All those formative years spent surreptitiously studying readers letters in *Woman's Own* (My husband loves to wear chiffon on the allotment - Desperate, Solihull) were at last about to bear fruit.

Our whispering caller said she didn't know the details, only that the chap's wife sounded extremely upset and very tearful and she hadn't liked to pry any more.

I didn't believe this 'reluctance to pry' rubbish for a second. After all she was a fully paid up member of the Personnel Mafia, whose motto was the same as the Special Boat Service, 'Not by Strength, but by Guile', with far more sinister overtones.

However, though calls of this type were unusual, I was happy to comply as we all lived in some trepidation of receiving a 'distress message' when away on duty.

Under normal circumstances, contact with home was in the other direction. We now had a telephone booth fitted under the stairs in the accommodation and available to the crew for two hours each evening. As the new service was constantly in demand by about 150 guys, it was imperative that you entered your name on a list, so that when the preceding caller's allotted time was up, the Radio Operator would tannoy for the next customer.

This meant that at any one time there would be about half a dozen blokes hovering in the vicinity, each fearful of missing their call. Convincing themselves that the swine on the phone had been speaking for much longer than his allotted six minutes alleviated their boredom.

This false conclusion would of course create discontent in those at the back of the queue, culminating in a good deal of watch checking and synchronisation. As the short lived horological exercise achieved nothing, people resorted to the tried and tested method of stress relief by shouting for the caller to, 'bloody well hang up'.

It wasn't only Einstein who believed that time was relative.

I digress; Tom duly arrived in my office expecting to discuss his annual report, (another farce developed by our friends in personnel) only to have me break the news that he was to contact his wife, as she had been in touch with the office and needed to speak to him on a matter of some delicacy.

Naturally he looked somewhat shaken and said he hoped it wasn't anything to do with the children.

I did my caring act and sat him at my desk, saying I would wait outside until he had finished. Unfortunately, he asked me to stay, something that I must confess I wasn't looking forward to. I rather preferred the idea of being a 'caring person' from a safe distance.

Anyway he seemed keen, so I sat down and prepared not to listen. However, the part of the one sided conversation I could hear soon began to fill me with a deep sense of foreboding.

'Hello, is that you Madge, it's Tom here. What's the matter, is it the kids?'

Pause.

'What? I know we've had some difficulties but I said we would talk about it when I come home.'

Pause.

'I don't care; I don't want that bastard coming around when I am away.'

Pause.

'Look, if you just give me a chance I know we can sort it out. Your mother doesn't need to know.'

Pause.

'For god's sake don't cry. We aren't the first people that this has happened to; it's not the end of the world.'

Pause.

'I am not just saying it. I know we can sort it out, but I tell you if that sod comes near I'll kill him.'

Pause.

'Of course I do, look I'll have to go now. Don't worry it's not the end of the world.'

Pause.

'Take yourself and the kids around to your sister's; she'll sort something out till I get back.'

Pause.

'Of course I do, we can fix it but I have to go now. See you next week.'

And he put the phone down.

As the conversation went on I became more and more embarrassed, convinced that, despite Tom's plea's to save the situation, I was witnessing a complete marital breakdown.

I also realised that the Project Management courses I had attended had failed to include a session on counselling, something which I now intended to raise with the Training Department.

Not quite knowing how to begin, I said how sorry I was to hear that his marriage was failing and that maybe it was one of the disadvantages of working offshore and would he like a cup of tea?

Tom looked at me as though I was mad and said, 'What are you talking about?'

'Well, I couldn't help hearing you say we can sort it out and having the fellow next door keeping out of the way.'

'Do you want to know what she was on about?'

'Only if you feel you want to tell me.' (Yes please.)

'The bloody washing machine has gone on the blink again. I told her we could fix it when I got home, but she is in a panic about the kid's clothes. That's why I said go round to her sister's and use her sodding machine.'

'But what about the problem with the chap next door?'

'Oh him, he's just a bloody idiot, he works for the Council and the last time I was away she sent for him because the TV went on the blink. The daft sod managed to tip the TV over, break our wedding photographs and fuse all the lights in the house.

I told her if he ever came around again I would kill him.'

I sagged back, partly disappointed and partly relieved that I didn't have a potential suicide (or murderer) on my hands. There is no doubt about it; Claire Raynor does these things much better than me.

Even so, I thought I had better give him a mild bollocking for having his wife waste everyone's time.

Tom said, 'Don't worry, when I get back I'll give her strict instructions not to ring again unless the bloody house burns down. Oh and could you do me a favour; don't mention this little problem to the lads.'

But the best bit came later when our whispering personnel woman called me back, obviously bursting to know what had happened. She began by asking a number of oblique questions like 'Are you giving him compassionate leave?' and, 'Is he terribly distressed?' 'Should I go around and comfort his wife?'

Talk about fishing!

I was able to reassure her that as the circumstances were of such a delicate and personal nature, I felt unable to betray his confidence. Furthermore I was convinced they both needed time and space to reflect on the situation and then determine the best way forward.

Absolute rubbish, but I enjoyed it immensely, it's not often you leave a personnel person speechless.

Someone Has To Do It

First of all I need to run through a recipe for making a 'smoothie' drink, something that's very popular in these days of healthy living and a desire to be anorexic. Now pay attention.

What we do is to cut up a selection of fruit such as apples, oranges, pears, pineapple and anything else available. We then push the cut pieces through the top of the mixer and onto the blades whizzing round in the base. They grind the fruit into a liquid, which flows under gravity into a jug and all that remains in the bottom of the unit is a mass of fairly dry pulp.

Later of course, you find that the remaining gunge will take no more than three or four hours to clean out in readiness for the next attempt to overdose on Vitamin C.

Now you may ask why I have come over all Nigel Slater. Well, it's because this is exactly how a Sewage Treatment Plant works on a platform. Although in this case, there are ingredients in the mix that it may be better not to itemise.

Disposal of sewage is like the cowboys in films; we of a delicate nature suspect they go somewhere in the desert when they require a 'comfort break,' but we never dare ask where or when. I guess knowing the truth wouldn't do much for John Wayne's image either, or maybe that's why he walked the way he did.

However, as I am determined to recount this tale, you need to understand a little more about this earthy subject if my little problem is to make sense. Be prepared..........

Sewage from the accommodation, together with graunched-up food waste from the canteen, is gravity fed into a large sewage tank located on the lower deck.

On entering the tank, the solids are ground up in a thing like a giant mincer. This is made up of a series of blades and cutters and nothing escapes its attention. The end result is that all solids have disintegrated into sludge and the macerated end product is discharged through a giant pipe deep into the sea. The munched solids are quickly dispersed, with most being consumed as organic foodstuff, which is why shellfish look so healthy. Okay so far?

As you might imagine, commissioning a sewage unit is not the most sought-after job in the world, but it has to be fully operational before people come aboard; either that or we have to take prunes off the menu.

One day, the news we all dreaded was conveyed to me by a caring operator; the sewage unit isn't working, there are 140 people on board, there is a smell you could strip paint with, this is an emergency and what are you going to do about it?

After I found where the lads were hiding, we fitted some temporary by-pass pipework from the tank into the sea, both in order to find out what was wrong and to keep the platform working (so to speak.) This also allowed time for the existing watery substance to drain off and let the sludge settle in the bottom of the tank.

You can see what a lot of fun we had that week, but it was still a step up from working in a mine, where there weren't any toilets at all (another little known snippet of social history for you.)

Tempting as it was to leave the temporary set-up in place and look for another job, we knew we were probably contravening several 'pollution of the sea' type regulations and so there was no alternative but to fix the machine, or preferably—go on holiday.

We had to get inside the tank to see why the 'mincer' had stopped working; but sadly, it was buried under a whole lot of rapidly setting sludge, which would have to be removed. The big question was, by whom?

My so-called team had already told me in words of one syllable (words like 'sludge' only more Anglo-Saxon) that they would resign rather than try to resolve this fascinating and unique engineering problem. At least that's the way I tried unsuccessfully, to sell the involvement idea.

So, in some desperation I contacted my Engineering guy onshore and having explained the dilemma (staff revolt and mass sacking versus a platform shut-down) he said 'leave it with him'— he would try to find help.

A day or two later he sent me a telex saying he had found a small team who were willing to clean out the tank and that they would be travelling from Hull in the morning.

We then spent a nice couple of hours trying to figure out just who would be idiotic enough to volunteer for such a job. One chap rather cruelly offered the opinion that cleaning out sludge must be better than staying in Hull. I didn't care, all I needed was a solution and I looked forward to meeting these rare creatures the next day.

As it turned out, there were three of them, a Foreman and two assistants. They all had one thing in common; they had no offshore experience whatsoever.

What they did have however, was considerable experience in cleaning out (by hand) ships bilges and as a sideline, scraping barnacles off the bottoms of fishing vessels.

Now, on a scale of desirable jobs, bilge and sewage tank cleaning ranks someway above that of a Tabloid Newspaper Editor. However, on a scale of essential jobs, it ranks just about level with that of a Brain Surgeon. If you think that's an exaggeration, the next time you're feeling bored, try asking a Brain Surgeon to remove solidified sewage from the bottom of a large tank.

My prayers were about to be answered.

I dressed them in the appropriate safety gear and led them to the scene of the crime followed, I noticed, by my loyal team at a discreet distance. We inspected the site, me from the outside and the foreman from the inside after which he said the magic words 'No problem Boss;' he would put the lads to work straight away.

We had already rigged up lights and vented the tank to make sure there was no accumulation of harmful gases and, just to make sure, we were testing the air at regular intervals. I wasn't about to lose my new-found friends.

Bear in mind that the phrase 'lack of harmful gases' should not be interpreted to mean there was no smell. There was; and it was about to get a good deal stronger and more widely distributed, as the lads disturbed the crust that had formed on the surface, by poking at it with their shovels.

Anyway they began to get stuck in (literally), filling buckets and clearing a path into the horrible darkness, coming out only to empty the hobgoblins over the side.

Later that day I paid a visit to the site and asked the Foreman how things were going. He said there were no problems and the tank was being cleaned out okay. On querying how long his lads had spent inside the tank, he said they had been in for about four hours so far.

He then added something that summed up the meaning of true dedication to a job that no one wanted.

'Lenny's okay, but George has to come out now and again to be sick.'

The funny thing is that word of the job this little trio were doing began to spread and soon numerous workers could be seen leaning over railings watching in fascination as the lads pulled out bucket after bucket of undercooked fertiliser and tipped it over the side. Less curious was that the appreciative audience carefully stayed on the windward side.

Enough detail, suffice it to say that thanks to the efforts of the lads from Hull, we had the tank cleaned out in three days. We were able to repair the damage and put the sewage system back to work, much I guess, to the delight of the local mollusc population who must have been baffled by the sudden lack of ready prepared meals.

How else could they become big enough to make Scampi and Chips, that 'must' for the enlightened diner. I suppose it's something to do with us all being a part of nature's food chain—ask David Attenborough.

Everyone was greatly impressed by the willingness of the trio to tackle such a crappy job, even though there were a few nervous tremors when they arrived in the Rec room one evening for a game of snooker. One of the stewards cautiously approached and asked them what they had done with their working clothes; I think he had a nightmare vision of having to run them through his washing machine. The Foreman assured him that all was taken care of and the somewhat stained and smelly overalls had been consigned to the incinerator.

Realising they had earned our undying gratitude, the Foreman approached me later and asked if there was anything else they could do whilst they were on board.

Contractors actually volunteering? It was like a dream. I was so taken with the offer that I purloined (borrowed) three sets of Company overalls and gave them to him as a thank-you present. As you can see, I certainly knew how to put other people's money to good use; I guess it's a gift.

Contracting the lads was no problem; in fact it was a golden opportunity (I never did get to apologise to that idiot consultant.) As there were any numbers of cleaning jobs to do, we ended up using the team for several months, after which the platform was gleaming. They were proud of their efforts and my one dread throughout their stay was that the Drilling Department would get up to their old tricks and cover the place with mud.

Best of all from my point of view was that as a show of gratitude, I used to receive a parcel from Hull each Christmas containing two superb bottles of wine. Obviously I had to wrestle both with my conscience and my duty to our Company regarding the need to declare these unsolicited gifts. So, I did the only thing a person in my position could do, I said sod it and we enjoyed the wine.

26

And Finally....

As I said at the beginning, it wasn't my intention to write a definitive account of offshore life; there are enough 'learned' works supposedly dealing with this subject already. I notice however, that the majority of these dissertations seem to delight in concentrating on the perceived risks (fire, explosion, blow-out, bad weather, travelling by helicopter, etc.)

In my opinion, there's been a marked tendency to overly emphasise the grittiness, stress, misery and marital discord caused as a result of working offshore, something which is now commonly referred to, would you believe, as the 'Psychosocial Aspects'.

If the experts who invent these phrases think that working offshore is overly stressful, it's a good job they weren't around when we worked three miles underground. Still, I suppose we didn't have to worry about bad weather.

I think it's been convenient for the psycho-babble merchants to somehow ignore the fact that working offshore wasn't compulsory; it's not like being in the Army, you could leave at any time and you weren't even being shot at.

Sadly, in our ignorance of psychosocial thingy, we thought that stress and fatigue was something you calculated when designing structural steelwork.

Scientists have, at great cost, developed a theory for why time flies when you are having fun. Apparently they believe that if the brain is concentrating on many aspects of a task, then it has to spread its resources thinly and pays less heed to time passing.

They should have asked us, we could have told them that as far as we were concerned there weren't enough hours in the day, even though there were times we wished we could be elsewhere or preferably, in bed.

Looking back, there is no doubt we had a good deal of fun. Somebody once asked me how I remained so cheerful and thinking about it, I could only reply that it was because I didn't fully understand the seriousness of the situation. This may be a failing on my part, but hey, it worked for me.

I read somewhere that the ancient Greeks and Romans believed that laughter and humour were perverse and degenerate and that people exhibiting these strange traits were suffering from an unfortunate lack of self-knowledge. I bet their Project Meetings were a whole *lot* of fun.

But there again, Rome didn't build the Empire by holding meetings, but by slaughtering anyone who opposed it, which, I guess, was an early example of the Italian management style, later adapted to great effect by both the Mafia—and a number of offshore Project Managers.

Obviously there were occasions when having 'fun' in any shape or form was unthinkable. Making the distinction is, I believe, largely dependent on the situation at the time. For example, during the episode of the bomb on board, a liberal use of 'gallows humour' to reduce stress and to make the time pass quickly was certainly beneficial and in no way detracted from the work in hand.

The key here is that we weren't really convinced there was a bomb, although I can't say how we arrived at that conclusion. And again, no one was hurt.

Conversely, I lost a very good friend who was taking part in a Search and Rescue exercise as a volunteer, when for some unaccountable reason the helicopter crashed into the sea with the loss of life of all on board. There was nothing to laugh about that day.

I don't subscribe to the view that 'accidents will happen'. However, I do believe that hazardous occupations increase the likelihood that an untoward event may take place. Vigilance is sometimes jeopardised due to tiredness or lapses in concentration.

I worked for many years in the British mining industry, which had the best deep mine safety record in the world. And yet an ex-student of mine, who had a fine career ahead of him, was killed in a rare underground explosion.

So yes, there was sadness on occasions and a time to grieve, but that seems to me to reflect life in general.

Overall I wouldn't have missed my offshore adventure for the world, partly due to being involved in tremendous technical and logistical challenges. But most of all, it was because of the pleasure of working with a group of people who could accomplish the most daunting tasks whilst retaining a manic sense of humour. I have always enjoyed working with like-minded simpletons.

Confirmation that I was not alone in having enjoyed our time together was brought home to me recently when I attended an excellent

Anniversary Dinner organised by the Company. We were watching a film showing the development of our largest Platform and I was coming over all nostalgic when Harry (you remember - our 'organised' Project Manager) leaned towards me and said, 'Bloody hell Brian, did we really build that monster?'

Looking at the pictures showing the Platform taking shape, I realised what he meant, but could only say, 'Yes, I guess we did.' A bit of a lame reply that didn't really do justice to what we both felt.

Coincidentally, at the end of the film, Walter the OIM, who had been providing some narration, summed it all up by saying, 'It was a wonderful experience and given the chance, I would do it all again.'

I couldn't have put it better myself—in fact all I need to do is to change my elastic-waisted trousers for elastic-waisted jeans and I'll be ready to go.

Any errors, omissions or chronological problems in my tale are the fault of a failing memory; it's hard to be nostalgic when you can't remember who did what to whom or why. One thing I can promise, however, is that all of the 'incidents' are absolutely true; it would be a disservice to my colleagues to have invented either the situations or their involvement in them. The only liberty I have taken with the truth is to have amalgamated a number of characters and I changed the chronology of some events.

I believe it was Sam Goldwyn who, when describing the kind of film he wanted, said, 'What I want is a story that starts with an earthquake and builds to a climax.'

I think that's just what we experienced during the great era of offshore development.

My thanks to you all for being there with me at the right time. Writing it down has at least made me smile; I hope others may experience the same pleasure.

The End

Biography

Brian Page is tall for his age; he was born and brought up in Knotty Ash and worked in Liverpool for a huge engineering company which has since disappeared. He then migrated into the Lancashire coal-fields, which seemed a good idea at the time. Later on he became a Lecturer in Mining Engineering before venturing even further north to Aberdeen.

Having adjusted to the daylight, he worked offshore in various capacities throughout the seventies and early eighties, before being brought onshore to do other things. Now retired, he divides his time between consultancy work and remembering the past.

Brian has been happily married for forty-six years to the same lady and a long time ago they produced a son and two daughters. He has yet to be recognised as an international 'man of letters', but in the meantime is determined ignore the facts and live forever.

Also available from PlashMill Press:

Poaching the River by Rod Fleming
ISBN: 0-9554535-0-X   Price: £11.99
A hilarious tale of life and love in a Scottish fishing village.

The Tobacconist by Jennifer Dalmaine
ISBN: 978-0-9554535-2-6   Price £9.99
Mystery and suspense in this thriller set in old Aberdeen.

A Boy's Own Mining Adventure by Brian Pqge
ISBN: 978-0-9554535-3-3   Price £11.99
Brian Page goes underground in this funny and touching book.

Silver Threads by Tom Ralston
ISBN: 978-0-9554535-5-7   Price £9.99
Love and drama set in the Scottish fishing industry.

Park Street Onwards by Sid Robertson
ISBN: 978-0-9554535-4-0   Price £11.99
The story of Aberdeen from the 1950's through the eyes of one man

All these titles are available through good retailers and online.

Alternatively, please contact us through our website at www.
plashmillpress.com

You may also contact us direct at at: PlashMill Press, The Plash Mill, Friockheim DD11 4SH, Scotland. Tel: 01241 826108. UK carrriage is free for direct sales, please see the website listing for international carriage rates.